Text Copyright © 2020
by AFOUGHAL MOHAMED
All Rights Reserved

Everything Is a Computation.

Author: Mohamed Afoughal.

Translated by: Abdelali Zaour.

To contact the author:

Email:
MIHMIRAW@HOTMAIL.COM
Phone: +2126.50.53.09.35

To contact the translator:

Email: Abdel63811@gmail.com

IMPORTANT CLARIFICATION BEFORE READING THE BOOK:

I would like to point out that in this book some scientific theses may be mentioned, or the names of some researchers and doctors, or some names of books and films, published in accordance with the right to share scientific information. All Rights are reserved, and I also would like to clarify that I do not consider myself a writer, writing is a skill, a profession, and an art. As a mathematics professor, all I master is arithmetic and mathematics. I have ideas and information, and I thought that it would be great to reach people so that I might change their way of thinking to the best. I used several methods before writing this book, such as talking with people, discussing and persuading them, and their reaction was positively good, which made me think of writing this book, which I think that it is a tool that I can use to spread my ideas to the public, so I would like you to not expect a prominent literary style in the book. The method is simple and the important thing is to communicate my thoughts to the reader in a simpler way.

Copyright © 2020 AFOUGHAL MOHAMED All Rights Reserved

Acknowledgments

To my mother, father, and brothers.

To all my friends, neighbors, and everyone I shared with whom a house ... a street ... a neighborhood ... a city ... a school. ... a section... a country ... a planet.

To everyone who taught and guided me.

To all the teachers who taught me, and thanks to them I was able to write this book.

To everyone who helped and encouraged me to complete this book, even if only by kind words.

To everyone who believed in me and gave me the energy to carry on.

To everyone with whom I shared my ideas and inspired me.

To every person on the planet who contributes even a little to improve humanity.

To all the scholars, writers, poets, and painters who left us with this momentum of knowledge.

... Quotes about the book ...

"Science has explained many things around us, so the path to knowledge begins with learning."

"My words are not a law which I impose on you, but rather a plow with which the minds are plowed to reap new, better, and more general ideas. I know that I will not be able to reach the truth by myself, even if I reached its depth."

--Mohamed Afoughal--

CONTENTS

Introduction..8

Computation..15

The Human Mind and Artificial Intelligence 26

Human..38

Universe..78

The Past, Present, and The Future...............106

Death...126

God and Computation................................134

Conclusion..144

Scientific concepts and theories....................149

References...181

Index...182

...INTRODUCTION...

" We keep moving forward, opening new doors, and doing new things, because we're curious and curiosity keeps leading usdown new paths."

Walt Disney

Everything Is a Computation

When I was a child, the radio seemed strange to me. I used to think that there is a genie or goblin inside of it, talking about the news of people and changing voices and languages. When I turned 13 years old exactly, the television device was about to be in every house, so people started forgetting about the radio at that point, I was lucky because my mom gave me the radio that we had to play with since we do not use it anymore, I was curious about it, I always imagined what is inside of that box, I asked my mother if I can dismantle it and she said yes, you can do it, just do not make a mess. She would have done anything to see the smile on my face, my mother, I appreciate everything you have done for me. I love you, mom.

Even though I was a kid, I knew how to use tools, I was able to dismantle the radio without causing a scratch on it, though the shock was big, I found no goblin and no genie as I had imagined. I only found lamps, wires, and other parts, which I knew nothing about what their roles are at the time, but I was sure that the sprites were hiding inside them, so I imagined that the genie lived inside the condenser, yes, it was my curious little mind.

I got upset a lot because I could not find anything extraordinary inside of the radio. As a result, that enthusiasm and curiosity disappeared, then I collected the debris in a bag and gave it to my mother to throw it. Despite all that, I was trying to guess how it works, but I did not find figure it out at that time, I waited four years to reach the baccalaureate, so that I would know the logic, mathematics, and physical relations and know more about light and sound waves. Then came the articular lesson, which I consider one of the best lessons I studied in high school. It was about electromagnetic waves and transmission., for I learned much about it.

The first thing we studied about it is how a person's voice gets broadcasted in different cities in the world while that person remains in one place. I learned how the sound turns into an electromagnetic wave that travels through space to reach a house, and then the house owner determines its frequency on the radio to pick it up, then it turns into electricity that passes through the wires to reach the amplifier and return to its original form.

I learned how an electromagnetic wave could include sounds, and I also learned a lot about the radio in general.

Learning about the radio popped up a whole lot more questions, the answers to which became possible, as I finally realized that all I needed to answer these questions was knowledge and more learning.

I had many questions that I could answer by learning, among which, how does the image work? I read articles on optics and light, and then the answer was that an image is only a group of small lamps that produce light from the separated photons from the electrons passing through the electrical wires.

Photons are the substance of an image, this answer, however, did not convince me a bit, so things were ambiguous for me, so other questions started arising, How do the colors change according to each angle in an image? How can photons be organized on TV or phone to complete the image? The answer was that the colors change according to an existing relationship between the optical wavelength and other electrical constants. Each color is a light, and the light is an electromagnetic wave, and the wave is related to a relationship with electrical variables such as intensity and tension, etc. The appearance of a single color is sufficient to control the variables.

However, there was still a mystery about how the colors are organized and arranged on the screen to look like the real shape of an image. I uploaded a picture of myself in the pain program on the computer. (You also can try that) Then I zoomed the image on the paint program, and I found small squares, each square takes only one color while the picture surprisingly looks real and its colors are organized.

I was finally able to break that mystery, as I became informed with the techniques of small squares called Pixels that organize these colors so that the image looks closer to reality. The image is divided into small squares, and each square takes its color, and when the squares meet, they make up the image that finally makes up the original. I learned then that the more the image was divided into a large number of squares, the closer it would be to reality.

I was able to find convincing answers to all the questions that I faced at the beginning of my knowledge journey. However, knowledge is a deep ocean, the deeper you dive, the more questions emerge, and whenever you answer one of which, several other questions pop up, and so was the beginning towards knowledge where I dedicated my time to find answers for many questions I always had, among which, who are we? How do we use our abilities and skills? How does the universe work? What is death? How can we determine our futures? Where does the past exist? Whenever I decided to search for an answer to a question I had, I found an answer with a mathematical dimension, and I began to realize that everything is a calculation that science can explain. So I decided to share with you all the ideas and the answers that I have reached in my journey towards recognizing computation by publishing this book.

...COMPUTATION...

" Fly flight is just a great phenomenon to study. It has everything - from the most sophisticated sensory biology; really, really interesting physics; really interesting muscle physiology; really interesting neural computations."

Michael Dickinson

Computation is the process that you do unconsciously, and it happens around you without you noticing, For example, when you read these lines, your mind is calculating and making complex computations, yes, you make sense of letters like a machine, and you try to understand each word separately so that you can understand the meaning of the whole line without realizing how you did that. Reading is not the only one that depends on computation but everything else in our lives, including speech, Because when you speak, you are not thinking about the process of respiration, it is done automatically, we do not think about it, but it requires complex and precise calculations, then the letters are pronounced at different frequencies as if you were programmed on that. Words come out of your mouth balanced and identical to your language, you do not know that that oxygen, which you inhale is necessary to complete the speech process, the muscles responsible for speech need oxygen to expand and contract, allowing you to pronounce in a specific tone. Each letter has its own output and has its own frequency.

This is how computation is manifested that your voice strings do without you noticing. Are you beginning to notice how much computation is close to us and how it an unconscious process that we do all the time? There are too many examples, I shall discuss some of them in the upcoming parts of this book, but now we will try to know what is meant by calculation or computation, and to grasp a definition of the term, we will start by answering the following issue:

How do we prove the existence of the computation using computation? For example, how do you prove that you are speaking according to software and mathematical mechanisms while you are talking? Using speech to explain how to speak, the answer to the issue is not complicated. Perhaps it must be understood well until we reach a convincing answer to it, so we will answer it in a scientific, mathematical way.

We all know the commutative law of addition, it was the very first we all studied in primary school, do not worry if you are not familiar with it, I will share it with you again.

According to the commutative law of addition:

{If you take a number and add another number to it, you will not get "water" but rather a fixed number}

For example, if 1 + 1 = 2. Then the form 1 + 1 = 3 is false because the sum of two numbers is a constant number, and 3 is not equal to 2. Also, 1 + 1 = water is false, because water does not belong to the numbers, this is how commutative law of addition works. We have demonstrated the mathematical calculation (1 + 1 = 2), with the mathematical calculation (plural is an internal syntax for numbers).

There are many mathematical examples in which the proof of an arithmetic object is performed with mathematical operations (for example, the proof that an arithmetic equation accepts a solution through consecutive and equivalence calculations, etc.)

We have now solved the problem (How do we prove the computation using computation) by analyzing some examples from mathematics.

We will try to answer it in another way so that you understand clearly, keep in mind that the main point of this is to reach the intended meaning of computation in the title of the book.

Soil law, Have you ever heard of a law with this name? If the answer is no, I also have not heard of it, and if there is a law called the soil law, then it is not what I mean, and if it is no law of soil out there, you'll learn about it in this book.

According to the law of soil:

{If you plant a seed of a fruit or a particular plant in the soil, the tree of water will not grow for you, but rather a tree or plant will give you a certain number of that seed}

For example, if you plant an apple seed, grow an apple tree, not a water tree.

Do you know why the water tree will not grow? The soil law is already in place; The soil respects it, and it cannot be broken.

Everything Is a Computation

You may think that it is evident that an apple tree grows if you plant an apple, but ask yourself why and how? Or maybe you are still subject to the calculation process, you do not want to understand it, you will find that the soil law is self-evident and does not require proof, so I will give you another law to free yourself from restrictions so that computation becomes more evident to you.

The law of stone, in order to avoid confusion, this law exists in this book only,

According to the stone law:

"If you add one stone to another ... without breaking them or adding them to other materials, you will not get "water," but you will get two stones."

Example: Lift a stone from the ground and put it together with another stone, how much will you have? Two stones, of course.

Now, did you ask yourself why two stones? And not water?, because, of course, it is stated in the text of the Stone Law. Have you noticed the importance of the Law of Soil after you got to know the Stone Law?

If you understand these laws and feel the importance of their interpretation, know that you are starting to break free from the computation and that you are starting to know what it is, these laws are just calculations, and it is only one aspect of commutative law of addition. Do you remember it? It is the first law in which we started our journey towards finding an answer to the problem (How do we prove the computation using computation).

If you can link the law of the soil, the law of stone, and the commutative law of addition, you are on the right track towards knowledge, and if you cannot link them, there is no problem, just go back to the text of each law and compare the word "water" between them and you will understand. The soil law and the stone law are an extension of the commutative law of addition - they are nothing but a set of calculations that we can prove with the computation, for example, to prove the soil law, all you have to do is to sow an apple seed and wait until the apple tree grows, then the law is fulfilled.

Since there is no opposite, then the law is valid. The cultivation process is a chemical calculation in which the mineral substances of the soil and the seed planted interact according to chemical equations with a mathematical dimension; therefore, we have demonstrated the existence of the computation (soil law) with the computation (cultivation). The same thing also goes with the stone law, where we have applied the commutative law of addition on stones. Have you seen how we were able to prove the existence of the computation using computation? Have you started to see the bigger picture here? Do you realize what is meant by the computation in the title of the book? ... Well, I think you do, but you cannot quite express it, even the expression itself is a computation.

Do not worry; the same thing happened to me when I reached this stage of thinking. All that matters now is that you realized it and started to distinguish it.

The title of the book defines computation as: "the logical and equivalence sequence arranged based on mathematical laws, just like in mathematics." The calculation is applied to everything, as it applies to numbers and functions (for example, arithmetic laws, addition, multiplication, etc.) So we can understand everything with the calculation; in other words, everything is a logical valence sequence arranged based on computation laws.

Everything Is a Computation

... THE HUMAN MIND AND ARTIFICIAL INTELLIGENCE ...

"Nobody phrases it this way, but I think that artificial intelligence is almost a humanities discipline. It's really an attempt to understand human intelligence and human cognition."

Sebastian Thrun

Everything Is a Computation

The mind is, linguistically, related to reason and rationality, and, conventionally, it is the mind that sets us apart from the rest of the creatures, for it is the mind that makes us differentiate between right and wrong, good and evil, and between permissible and forbidden. This is how the mind id defined, so how do science and computations define the mind?

As I have already repeatedly mentioned that everything is a computation, the mind as well is nothing but a set of mathematical calculations and algorithms, in order to understand them we must dig deeper and go as far as possible; Everything is a logical equivalence sequence, so if you want to understand one thing, suffice to associate it with another thing in a logical sequence then if you understand that thing you will understand the other thing. Just as humans did when they used computation to understand how does rain work, so they linked rain with the presence of clouds and linked the clouds with seas and seas with water vapor and water vapor with temperature.

Humans tried to study rains extensions, such as heat, moisture, etc. so that they can understand how does rain works. We will do the same thing with the mind to understand how it works by studying some of its extensions and explaining how it operates.

Artificial intelligence is an extension of the human mind, just as the painting that the artist paints is an extension of his or her mind; the lines the writer writes are an extension of his or her mind. And the speech the speaker formulates is an extension of his or her mind, suffice for you to look at the painting, read the written lines, or listen to the speech to understand all the thoughts and opinions circulating in their mind.

Therefore, we can say that to understand the human mind, it is sufficient to understand artificial intelligence, where there is a logical sequence between the two, as between rain and temperature. This is why I named this axis the human mind and artificial intelligence.

Almost a week does not pass without hearing about a scientific breakthrough in the field of artificial intelligence technologies, such as Google's AI software which is capable of conducting philosophical and artistic dialogues with humans just like the human mind ... The victory of the program that Google developed under the name (Alpha GO) over the world championship in the game Go in four out of five rounds.

A victory that was a challenge in front of artificial intelligence scientists and most optimists argued that it will not happen before another ten years. Likewise, the success of a short story written by a Japanese artificial intelligence program in passing the first stage of the monthly literary competition qualifiers, and this strongly indicates that the artificial intelligence is an extension of the human mind, as it deceived the writers who did not realize that the story is a product of computer program only after the competition ended. All this indicates that artificial intelligence is only another facet of the human mind, so what is the history of artificial intelligence and how it began and developed?

The term Artificial Intelligence was coined by computer science researcher Allen Newell in 1956. The 1950s is considered the birth of computational artificial intelligence science as we know it today.

However, the idea that there was (or created) a subjective mechanism in the human mind occurred in a period much earlier than this, older than you can imagine. In ancient Greek mythology, one of the myths tells about Hephaestus.

The son of Zeus, the king of the gods and Hera, his queen. He was ugly, so he was cast from the sky to the earth to become the god of fire and the blacksmith.

The legend tells about Hephaestus, who made gold and silver statues that can move and defend god's places and defeat enemies. This made-to-measure statue was the first imagination of a mechanism capable of acting on their own, as Homer mentioned in his poems.

Greek mythology also tells of Talos, a copper giant made by Hephaestus - and possibly Daedalus - at the request of the god Zeus to protect Europa on Crete, Talos roamed the shores of the island three times daily to protect it from pirates and invaders; doesn't this remind you of the gigantic smart mechanisms in Terminator films? There are also indications of such ideas about smart artifacts and perhaps attempts to build them as well in ancient Egypt (Pharaonic civilization).

The Islamic civilization, more specifically the writings of Jabir bin Hayyan, we find that he discussed the idea of Genesis, which is an attempt to make inanimate sane or create a sane entity using the science of alchemy, but there is no evidence that he tried to do so.

Nowadays, we will find that the idea of an automated or industrialized entity possessing intelligence and being able to live and exercise different tasks and make decisions - even if they were terrible - was a recurring idea in Western literature and culture in general.

The history of artificial intelligence is closely related to the history of logic, arithmetic, and philosophy. Artificial intelligence presupposes the ability of the mechanism to think and make decisions in the same method and method of the human mind, which is something that philosophers spent decades trying to understand and establish general lines for, starting from Plato, Aristotle, Euclid, and William Okami, as well as many Arab scholars such as Al-Khwarizmi, as well as René Descartes and Thomas Hobbes who believed that logic is only a calculation.

Likewise, Gottfried Leibnitz believes that any intellectual or philosophical disagreement between two people can be resolved and reached righteousness using a skilled calculation that simplifies the matter mathematically and provides the solution. In the mid-nineteenth century, scientist George Paul presented his theory of algebraic logic, later called Boolean algebra in his book The Mathematical (Analysis of Logic) and The Laws of Thought, a theory that relies on Only two values (zero and one) to represent any variables in any mathematical process, were then considered the basis for computer science and the nucleus on which the current information age was born.

Based on the compulsory logic of Gottlob Frege, the famous philosopher Bertrand Russell and Alfred North Whitehead wrote (Principia Mathematica), and he encouraged German philosopher and mathematician David Hilbert to try to solve the challenge at the time: Is it possible to generalize mathematical logic on all kinds of problems?

In what is known as the Hilbert program, and the solution requires success in making the mathematics language uniform and with intuitive and fixed rules; it is followed by everyone during the application of mathematical equations, so that they include all mathematical facts, with no contradiction or conflict between them, As well as the possibility of applying any mathematical equation on a real dilemma, reaching the right solution without the need to make idealistic assumptions that are different from reality, while creating clear logarithms that show the correctness or error of all scientific data.

They understood that any dilemma subject to mathematical logic could be automatically represented using the values 0 _ 1. Turing presented his theory that confirmed this (Turing machine), which is a theoretical model of a mechanism capable of processing any data using the binary system. Turing's work was the real beginning of the computer world, and his work then laid the foundation for artificial intelligence technologies.

History says that attempts to understand the logic of the human mind are seldom separated from attempts to understand the universe and the universal rules of thought, as well as the engineering of things, physics, and mathematics, which represent the basis for all branches of logic and that take on a mathematical form regardless of its essence, the human mind is subject to logic is, that is, by looking at its extensions, where we can express all the actions of the human mind, with logical algorithms and sequences, as it can be represented automatically using the values 0 _ 1 (zero and one are only symbols; zero means no, and one is yes) and this is what is shown to us in artificial intelligence.

The relationship between this representation of the human mind and what people perceive as a logical relationship, for people, the mind distinguishes between what is desired and rejected, just as it distinguishes between good and evil, just as a machine is programmed on that. So how does the mind program? Where is it, and how does it interact with the universe? All of these questions will be addressed in the next axis.

Everything Is a Computation

... *HUMAN* ...

"There is no physical law precluding particles from being organised in ways that perform even more advanced computations than the arrangements of particles in human brains"

Stephen Hawking

Everything Is a Computation

Since the dawn of humanity, humans look for solutions to problems. It has been necessary to find answers in order to survive. When the lightning struck and made a fire, humans wondered how the fire occurred, and knew that it was necessary for heating when it was cold. So they began learning how to make a spark to start a fire, even if it was small. They noticed when stones hit each other, it sparks, (electron fission), that spark was able to produce fire to benefit from it for many centuries in cooking and heating, without them having to ask for the scientific and mathematical interpretation of how that fire or that spark got produced, later on in time, we became aware of chemical reactions to conclude that the fire was nothing but a chemical reaction.

The fire is not the only one that our ancestors needed in order to survive. Instead, they needed many things. As humans evolved, viruses and dangerous diseases developed that threaten humanity, killing millions of them in just a few years.

Man has become more threatened than before, so if he needed fire to keep wild animals such as lions and bears out of the way, they now have guns to kill them, but the gun does not kill diseases and viruses, and given the severity that viruses and diseases posed to humans, they could develop medications and treatments for them. One of the groundbreaking things humans have achieved is a technique that fights the most dangerous viruses and dangerous diseases, but more than that; Preventing and avoiding it.

This technique is known as anti-serum or the technique of antibodies, where a person is vaccinated at a young age, for example, with a serum carrying a dead virus responsible for a deadly disease. Then the body makes antibodies to them, and then it stores it in the immunological memory, so if that person got infected with that virus in the future, the antibodies do their job and fight that virus, because now that person has immunity to it. All of this poses logical problems; how can the body identify the virus and create antibodies? How it stores it until it needs it? All this implies that the body is programmed, so how can this be done? Perhaps we must understand the human body using computation so that we can answer these problems.

Well, to understand the human body, we will imagine it as an electrical circuit and look for the role of each segment in that circuit, in every electrical circuit you find several parts, each piece performs its role, so that they are connected to each other with electrical conductors, for example in a computer the copper wire is the one that plays that role.

We will start by determining who plays the role of copper wire that conducts electricity, in the human body it is easy because it is enough to put your hand on a wire through which electricity passes to discover that the blood veins and arteries are connected to electricity. I do not advise you to try it, let us take advantage of the experience of the people who died after that to conclude that the human body's electrical conductivity is blood.

After we got to know the electrical conductor in the human body, and as known, in order for every electrical circuit be active, there must be an essential element in it, which is the electric generator, so who do you think plays the role of the generator in the human body? Is the liver, stomach, or kidneys? What do you think? Well, the whole abdomen plays the role of the generator because we need all its components to reach blood that conducts electricity, so when you eat an apple that grinds the stomach to extract mineral substances and materials thrown into the electricity industry, the kidneys filter fluids from unnecessary impurities that impede the electricity industry, such as a toxin, As the liver and spleen filter the final blood, it passes through the veins to the heart, so what is the role of the heart in this electrical circuit?

The heart is crucial for a person's electrical circuit to function optimally. The heart plays the role of the capacitor, the same role that the capacitor (condenser) plays in the radio, as it pumps electricity, that is, blood throughout the body, and with the help of what? Oxygen, of course, is the role of the lungs, where you receive oxygen from the outside to integrate it with the blood to send it to the heart, and it represents what is called the small blood circulation.

The condenser pumps electricity throughout the body, including the brain when blood reaches the brain, electrical calculations occur at the level of the brain. We must look at it because understanding it will help us understand the electrical structure of the human body.

Neurologists say: The brain contains complex electrical circuits, consisting of billions of nerve cells, and unlike the computer, the brain is made of thousands of cells that differentiate in shape and components, which communicate with each other and are distributed in several areas of the brain, and they also change in Brain conditions.

Each of these cells is a distinct electrical circuit. It receives information from thousands of neurons, analyzes it, and re-exports its electrical signals to and from each other. Neural signals may reach thousands of other cells, and this process takes only a fraction of a second, and this process occurs thousands of times per minute so that every one of the cells passes a hundred billion neurons over the life of each one of them.

Electroencephalogram (EEG) devices measure the electrical activity of the brain, which occurs as a result of an electric current flowing during the synaptic excitations of nerve cell branches, which are very sensitive to the effects of secondary currents, and EGG signals can quickly be recorded in a non-surgically electrode on the head through an electrode through the head. EGG devices are also characterized by low cost and portability, and they have a high time accuracy (up to approximately 50 milliseconds, meaning that the brain signals can be monitored every 50 milliseconds).

For all of these reasons, it is the most popular way to record brain activity. However, it gives relatively poor signals (since the signals captured by the electrodes must pass through the bones of the skull, scalp and several other layers), and they suffer from a high rate of disturbance (What happens either internally in the brain or externally on the scalp, or due to movement of the facial muscles, or eye movement), as well as decreased spatial precision. Magnetoencephalography (MEG) is a non-invasive method for recording magnetic brain activity using magnetic induction. MEG devices measure the flow of electrical currents through the branches inside the nerve cells, which produce magnetic fields that can be measured outside the head.

The neurophysiological processes that produce the MEG signals are identical to those that produce the EEG signals. However, while EEG devices are very sensitive to secondary sources of currents, MEG devices are more sensitive to the primary sources of currents.

The MEG devices are distinguished by monitoring the magnetic fields of the skull and scalp bones, so the interference in them is less than the EEG devices, and this indicates that the human brain produces electromagnetic fields precisely as the radio and phone do for remote communication and transmission, and this means that we can communicate from long distances without the need for a phone, and this is what some scientists call telepathy. Magnetic fields are detected by devices of superconducting quantum interfaces, which are very sensitive to magnetic strikes that occur due to the activity of neurons.

The electronic equipment used to measure the magnetic activity of the brain is cooled to about 273 ° C to facilitate the work of the superconducting sensors. The MEG devices require effective protection from the electromagnetic interfaces, the electronic equipment is installed inside a magnetic protection room so that the effects of magnetic fields from the external sources are mitigated.

In the late nineteenth century, scientists observed that the destruction of a specific part of the brain caused the same language deficiency in most patients. The destruction of the left frontal lobe in the Broca region - named after the French surgeon Pierre Paul Broca- damages the ability to speak, while it causes speech. The damage to the left temporal lobe in the Wernicke region - named after the German neuroscientist Karl Wernke - has difficulties understanding the language. These observations have led scientists to believe that the brain processes words in organized stages, through a series of language-related regions.

This is evidence that we do accurate calculations in the brain while speaking and talking without being aware of it. By using specific imaging techniques, such as positron emission tomography and functional magnetic resonance imaging, scientists can directly monitor the brain while speaking, listening, reading, or thinking.

Studies based on these techniques have shown that language processing is very complicated, as language regions occupy large areas of the brain, and these areas are activated in different ways, patterns, and different types of linguistic tasks.

Now that we know some of the mechanisms of how the brain functions, we will try to link it with the rest of the body's electrical circuit, and that link manifests itself in the regulation of the body's processes, where the main body's operations centers are located in the brain stem. While other brain stem areas regulate swallowing and stomach and intestinal movement, all this calculation occurs unconsciously.

In the hypothalamus, there are also neurological centers that control some of the body's operations, and most of these centers maintain the stability of the body's internal state. Some centers, for example, control the amount of water in the body, where specific neurons monitor changes in the level of water in the blood and tissues and transmit these the information goes to the hypothalamus, if the water level is low, the hypothalamus produces a feeling of thirst, which causes the individual to drink water. At the same time the hypothalamus sends messages to the kidneys to reduce the amount of water lost from the body, in case the water level in the body increases, the hypothalamus sends a message to eliminate the feeling of thirst and to increase the amount of water lost by the kidneys. As for the other centers in the hypothalamus, they regulate, under the same principle, the feeling of hunger and body temperature.

The hypothalamus is connected to the main gland in the body, the pituitary gland, through a thin body of tissue. The hypothalamus organizes many of the body's processes by controlling the production of the pituitary gland for chemical messages called hormones and their release. As for feelings and emotions, it participates in organizing the emotions that we experience in many areas of the brain and other parts of the body, and a group of structures in the brain called the limbic system plays a central role in the production of emotions. Emotion is usually aroused by an idea in the cerebral cortex or by messages from sensory organs. In both cases, the produced nerve impulses reach the limbic system, and the nerve impulses stimulate different areas of the system depending on the types of sensory messages or ideas.

Batches may activate, for example, parts of the system that produce pleasant sensations associated with emotions such as joy and love, and areas that produce unpleasant sensations associated with emotions such as anger and fear. This indicates that feeling is also a computational process performed by the human body.

As for thinking and remembering, the scientists have reached little information about the very complex processes that are related to thinking and remembering until now, but they agreed that this is nothing but mathematical and computational operations, thinking involves processing this information through circles in the area of the interconnected cortex and other parts of the brain, where These circuits enable the brain to associate the information stored in memory with information gathered by the senses.

Scientists are now in the process of a preliminary understanding of the simplest circuits of the brain. The formation of abstract ideas and the study of difficult topics require complex circles that require extensive studies. The brain's frontal lobes play a fundamental role in many processes of thinking that distinguish humans from animals, and they are especially crucial in experimental thinking, visualizing the expected results of the actions, and in understanding the feelings of the others.

Frontal lobes injury may lead to a loss of these capabilities and some aspects of human thinking such as religious and philosophical beliefs, which outside the scope of scientists' understanding now and may remain so for a long time. Scientists now have much to investigate about the organic basis of memory, some of the structures of the limbic system play significant roles in storing and retrieving information, and these structures include the amygdala and the hippocampus, both in the temporal lobe. Those injured in these structures may lose the ability to form new memories despite their ability to recover information related to the events that preceded the injury.

These people can learn new physical skills, but when performing these skills they forget that they did it before, according to some studies, memories may be formed by creating new brain circuits or changing existing ones. Both processes involve changes in synapses, meaning the structures where the batches pass from one neuron to another, glycoproteins and other large molecules control the synapses of these changes, and proving this general interpretation of memory formation requires extensive research to reveal the details of the processes associated with it.

The difficulty in understanding the mechanisms of the work of the human brain appears many despite the development witnessed by this field in our time, however, what scientists agree is that these mechanisms are complex computational operations.

All that we observe in this momentum of logical calculations and processes that the human body performs is that it is considered subconscious, meaning that the human body performs without awareness, and this is the reason, in my view, philosophers consider humans have a conscious part and an unconscious one, the outer mind, and the inner mind, all of these names hide a computational dimension behind them, for that the subconscious is the place where the mind calculates and processes without us knowing. You may realize that you are happy or sad, but do not control the release of hormones, all of this is done automatically.

Consciousness is a consequence of the unconscious, a result of an account we are unaware of. We have already discussed some of the mathematical examples that we do unconsciously in the brain, which leads to consciousness such as a feeling of hunger or thirst, a feeling of happiness or sadness, speech, and understanding of language.

I have read many books that discuss the uncanny powers of the subconscious and the paranormal subconscious, and what the inner mind can do, among the things that I became acquainted with by reading these books is that a human can change his habits and control his destiny by controlling his or her inner mind.

Computation not only corresponds to all of these ideas, but also supports them and adds splendor to them, for example in those books they tell you that you can master any language you want, or become an accomplished painter and even read other people's ideas and a lot of other things, all by programming your inner mind in many ways, they mention meditation, repetition, exercising, etc. How do we relate all this to what the computation interprets consciousness and the unconscious?

We all know that language, drawing, speech, etc., are nothing but skills that result from mathematical and computational operations that the body performs unconsciously, as we have already seen examples where we reached the fact that the subconscious is the sum of those complex unconscious calculations, so self-development and consciousness are inseparable from the development of these complex calculations, that is, the unconscious.

If you want to develop your linguistic software so that you can speak any language you want, you just have to repeat the output and reception of that language to find yourself programmed on it precisely as it happened with your native language, so your awareness becomes able to define it and understand it easily because your mathematical mechanisms are programmed on it, and if you want to to become a professional painter in the future, you must repeat the training to find yourself mastered the drawing skills. You will have many paintings just like other professionals, so if you would like to learn anything new, you must program your inner mind on it, just like the machine. That programming is done through repetitions, training, and exercises. This will make you master anything you want, just as it happens to the hand of a professional guitarist or a pianist. Believe those who said that the profession is refined with learning and mastered with practice.

Each one of us has a distinctive consciousness, and unique capabilities, each one of us interacts with factors according to which we were programmed, which means that all of us did not go through the same experience, nor we were exposed to the same factors and situations. And this is precisely what makes each person an entire universe distinguished of his or her awareness, separated by his or her own opinions.

This does not mean that you cannot program a person according to your awareness and beliefs or do the opposite. For that, there is an interactive logical computational chain that links us and that makes us learn from each other without being tied (computations that are done in the subconscious), we get programmed on habits and behaviors without even having to notice, it could be by reading books, watching movies, or even street signs.

You may find yourself asking for something from the shop without realizing it, due to publicity that you had seen somewhere without paying attention, you may become addicted to drugs because you watch a movie whose hero smokes weed, so you got inspired to do so just because the hero in the movie does it.

Films always try to market that there is a hero in the film, unlike what it has to do with reality, we are all equal, for me, there is no fictional hero in this life, you are only the hero, you must control your decisions and not make your destiny hostage what your computations receive from the outside.

Do you often ask yourself where is the hero in this life if we are all just programmed machines? And our brains are calculations that constantly program.

Program yourself to become better than your older self, make yourself a hero, learn a new skill every day, eating, and sleeping is a skill that everyone can master, learn something special, make yourself a unique person, with a great taste, always smiling, who has no difficulties, always analyzing and solving problems, always looking for solutions and not focusing on problems. You may be able to read other people's thoughts, but do not make it your concern to know what the other is thinking, in doing so, you make them the hero, and you put yourself in the place of the viewer, perhaps reading ideas must be a secondary skill that you should turn to when necessary.

For example, when you want to persuade a person of a request you want from him, first you must be convinced that this request will not be difficult for him so that your calculations pass a good signal to him that your request is simple, which will get into his mind smoothly, we have previously learned that speech is an extension of the human mind, so it is sufficient for you to listen to him to realize what he is saying, then put yourself in his place, and imagine that you are the one who is speaking, and you will find that your unconscious calculations are beginning to come close to his calculations. Thoughts will float on your mind when he talks, then decide how you will convince him, by persuading yourself, and after you embrace his thoughts that floated in your mind.

Imagine yourself sitting in a café with your boss, discussing a business-related matter, and you want to convince him or her to promote you and raise your salary, the first step as we previously mentioned is to first convince yourself, meaning that you are convinced that this request will not be difficult for him, therefore, try to start by showing the features that convinced you that you deserve to be promoted, without telling him about promoting you, then let him talk even if he speaks for a half an hour, put yourself in his place and repeat every idea he says in his mind until he finishes his words, then you will know how your boss thinks if it is a compliment, praising you then that is the opportunity to tell him that you deserve a promotion with a really persuasive family reason, in addition to practical reasons, for example, The work you do is highly requested by other companies with higher salaries.

And if he is a possessive person who boasts of all his thoughts indicating that he is praising himself, do not tell him that you want the promotion, just praise him and agree with him about his matter, and let him speak again, then listen to him again and repeat putting yourself in his place until you deduce that he cannot promote you. However, it can give you a grant for every job that you master, or increase your salary even a little, then offer him to do other tasks that cost him a lot in addition to your work, in exchange for a grant or an increase in salary, when the months pass. He is accustomed to granting you an increase in salary, then ask him more, or you will only go back to doing your job, then you will know your role and recognize your capabilities, so learn to share with people your capabilities and skills. You deserve more than sitting in an office and wearing a tie. You deserve more than you dream, do not let people achieve their dreams on the sweat of your forehead.

So the principle to read people's thoughts and convince them is to convince yourself of what you want first, then penetrate into their minds through its extensions such as speech and writing, etc., then put yourself in their place and try to convince them as if you want to convince yourself. This principle may also be applied to you by another person, just as a psychiatrist applies it to the patient to convince him, we always apply it without realizing it, and I just tried to simplify it by words and examples so that you can master it. And do not forget that everyone applies it, be good at it, and your opinion will make it clear and you will become wise with the computation.

We were able to demonstrate that a person's abilities and skills are only a complex computation that his body does. We have inferred this with the skill of drawing, speaking, and the ability to read ideas. As we have already discussed, hunger and thirst are nothing but calculations that the brain performs. From this, we can deduce that all human instincts are nothing but mechanical programs that are self-programmed in the brain.

The phrase that a person is nothing but a set of mere calculations performed by the body may appear obvious to some after all of that evidence that I have provided, as it may seem vague to some, so we will dive into the depths of the human body to make things unambiguous. We will do a thought experiment, in which we will create a robot, program it, study its behaviors, and compare them with our behaviors to see the computation that humans do unconsciously.

It's not necessary for us to go to a laboratory or factory to make this robot, we will only make it here using a pencil and paper, how is that! Do not be surprised, all you have to do is give me your undivided attention and imagine a robot that looks like a human in a room, an ordinary room in which there is a window overlooking the sun directly, and there are also two electric switches, and we are outside the room controlling the passage of electric current in the two keys.

Everything Is a Computation

On top of the robot, we put a plate to receive the sun's rays and convert it into electricity, and a thread from which to charge electricity also comes out, and there is a red button on the chest. We will charge the robot entirely until it's 100% after we program it to do something, for example, moving around in the room. When its charge capacity reaches 50%, it will try to charge itself by linking the thread to one of the two keys in the room. If one of the two keys does not work, it must pass to the other key. If neither of them works, it must direct its head to the window to pick up The sun's rays, but we program it that charging via the switch is faster than charging via solar energy, and when it is fully charged either with solar energy or through the two switches it will press the red button on its chest to empty all of that energy in it.

Now that we know the elements of our experience, we will give the robot a head start. Well, the robot started to move around the room as we programmed it until it reached half capacity, so it started looking for the first key, the first key does not work because we turned off the electricity from outside the room, so it tried to use the other, it does not work as well, so what will our robot do?

Everything Is a Computation

If you thought it was going to the window, you are wrong, have you forgotten? We have programmed that charging by electric current is faster than charging with solar energy, so it will try a third attempt in order to charge through the current, as the first and second attempts are inevitable because we programmed it, since it must add a third attempt because of our programming that charging with the current will be faster.

Here the robot is trying a third attempt, so we passed the current in it, the robot finally is fully charged, but it did not empty itself by pressing the red button, but a programming error has occurred because it did not understand how to empty itself, and it is programmed to charge it when its capacity reaches 50%, then our experiment ended.

Have you noticed how a robot could survive when it did not understand how to empty itself? And did you notice how it preferred charging the current only because we programmed it to do so? We also choose that delicious dish without realizing why we distinguished it from others, perhaps we chose it because it is full of materials that we need to make electric energy, so we may prefer bananas over the rest of the fruits without realizing that we chose it because it is distinguished by the energy we need.

Also, we may hate the smell of something without realizing that our calculation did not favor it because it is unhealthy or harmful such as the smell of bacteria, for example, the smell is only a chemical reaction at the level of the nose, the nose passes the different particles of those smells, which in turn stimulate the smell receptors to induce a specific chemical reaction, so the brain uses computation to know what kind that smell is, and thus the brain can distinguish the different and mixed smells to realize the type of those smells, so it distinguishes pleasant perfumes from bad smells from whatever source, and even distinguishes pleasant perfumes themselves in the degree of concentration or lightness according to the sources.

The sense of smell is an essential part of the respiratory system because the olfactory system in its basic design is similar to other sensory devices, there are cells capable of receiving stimuli, and neurons that transmit messages that say that they have received stimuli like these. Moreover, parts of the brain process the computational information contained in it subconsciously and turn it into a feeling and thoughts related to that feeling. The hair in our noses prevents the entry of dust and contains mucous liquid to moisturize the atmosphere, and all of these chemical reactions that occur are nothing but a chain of complex calculations.

All senses depend on logic and computation, where each sense can be explained by equations and logical correlations, for the five senses that a person possesses are the arithmetic windows that a person relates to external accounts. From this computation, there is an awareness of the external environment.

For example, to see an image of a scene, one needs eyesight. A vision or sight is the ability of the brain and eye to detect the electromagnetic wave of light to interpret the image of the visible horizon. The eye sees to distinguish colors and shapes and reveal the light from darkness, so when light passes through the lens of the eye, this leads to the reflection of the images seen on the retina, which in turn transmits the image to the brain that can perceive it.

Nevertheless, if a person wants to distinguish between the characteristics of things and recognize their properties, they will use a sense of touch. When the fingers come into contact with something, the skin acts as the conductor of the nerve endings, which in turn respond and transfer the perceived attributes to the brain to be interpreted and processed. Our somatosensory system distinguishes rough surfaces from smooth ones as it distinguishes the temperature of surfaces. All of this happens according to software and logical computations in our brain, as we have already mentioned.

According to our study and analysis of the computational dimension of the human senses, we have concluded that the human body is nothing but a set of complex computational calculations, as we realized that these senses have a limited extent that they cannot exceed. The practical reality has proven this issue since the five senses with their sophisticated mathematical devices are limited by certain limits and by specific capabilities that cannot be exceeded. For example, we cannot see what is currently happening in the streets of the city of Paris if we are in the city of Rabat, just as we cannot see what is happening in the street next to us and even behind the walls of our room, because the wall blocks light from the eyesight.

The sense of hearing also has specific vibrations that we can hear. If it exceeds this natural limit or decreases, we will hear nothing. In the ant world, for example, we cannot hear the ongoing dialogue between ants' despite our study of ants' kingdoms, armies, and systems, all because of the ability the ear's absorption of the sound wave vibrations is limited, and what applies to the sense of hearing also applies to the sense of smell and touch, that is, these senses have limited ability, so how can we feel all that is happening around us? How can we communicate with the outer universe that our senses cannot perceive? All of these questions we will try to answer in the next axis.

Everything Is a Computation

... UNIVERSE ...

" We're presently in the midst of a third intellectual revolution. The first came with Newton: the planets obey physical laws. The second came with Darwin: biology obeys genetic laws. In today's third revolution, were coming to realize that even minds and societies emerge from interacting laws that can be regarded as computations. Everything is a computation."

Rudy Rucker

"If you want to find the secrets of the universe, think in terms of energy, frequency and vibration."

Nikola Tesla

Everything Is a Computation

There are many aspects of computation in the universe, in fact, there are many phenomena in the universe that scientists have been able to transform them into mathematical formulas. for example, to demonstrate that there is a mathematical relationship between temperature and volume, the higher the temperature of the ball, the greater its size, and because of the density we demonstrate who will float on the surface of the water, Archimedes used water and objects to discover this, and he managed to convert all of this into a mathematical equation between mass and volume, Albert Einstein was also able to link mass, energy, and speed with a mathematical formula.

There are many examples that indicate this. Several theories have emerged that discuss the possibility of computerization of the universe, meaning that the universe is a huge quantum computer, so what is computing? This has always been a pivotal question in the field of computer science.

At the beginning of 1930, computing meant the function of people who used to operate supercomputers. And at the end of 1940, computing was defined as: a set of steps implemented by automatic computers to produce knowledge outputs, and this standard definition remained for fifty years after that, but now it faces many challenges; As people from many fields began to accept the idea that computational reasoning is a way to understand science and engineering.

The Internet is full of services that do computing without stopping. Researchers in the fields of physics and biology claim to discover computing processes from nature unrelated to computers, computing is now in all sciences. Computer scientists design, build, and program computers. But again we go back to our question, what can be considered a computer? What is the rock missing (like a physical system) and owned by a computer?

According to Seth Lloyd, author of Programming the Universe, all the interactions that occur between molecules in the universe not only transmit energy, but also information, in other words, molecules not only collide, but perform mathematical operations.

The entire universe computes, each An atom, an electron, and an elementary particle that keeps a lot of information, and every time two molecules collide, these two bits are processed.

By delving into the computing power of the universe, we can build quantum computers that store and process information at the level of atoms and electrons. This computing power forms the basis of complex systems and provides a deeper understanding of the origin and future of life.

Computing is taught in two ways: theoretically and materially; Mathematics is concerned with studying the theoretical side of computing, by providing mathematical definitions of computational matters such as algorithms, and providing theories about their properties.

The most crucial idea in computation is digital computing, which was portrayed by Alan Turing, Kurt Gödel, Alonzo Church, Emil Post, and Stephen Kleene in 1930, where they discussed the basics of mathematics. One of the most important questions was: is first-order logic decidable? That is, is there an algorithm that can decide whether a particular logical phrase (first-order) is a theory? Turing and Church proved that the answer is negative, there is no algorithm with this description.

To demonstrate this, they provided an accurate description of the unknown concept of a computing function and writing it in an algorithm. Turing did this through what is known as a Turing machine, which is a virtual machine that processes separate symbols written on a tape compatible with a limited number of commands. The study of computing functions is made possible by the work of Turing and others. According to the Turing-Church hypothesis, any function that is intuitively computing is computing with the Turing machine, this can be formulated as follows: "Any function that is seen as naturally computing is a Turing-computable function."

Intuitively computable, which means that it is computable by following an algorithm or an effective procedure. The effective procedure includes an expired list of clear commands for the production of new symbolic structures based on ancient symbolic structures.

Some scholars study that the physical universe is a computational basis, the universe itself is a computational system, and everything inside it is a computational system as well, where it is viewed according to two different perspectives, the first of which is an automatic operator that represents the traditional computational model and the quantitative, non-quantitative.

The idea that the universe could be a giant digital computer existed decades ago. In 1960, Edward Fredkin, then Professor of the MIT Institute Konrad Zuse who built the first electronic digital computer in Germany at the beginning of 1940, proposed the idea that the universe is an integrated digital computer (and recently the idea was supported by computer scientist Stephen Wolfram.

According to these physicists, the universe is a giant cellular robotic operator. The robot is a network of cells, with each cell taking a state from within a finite set of states and updating its state with separate steps according to the adjacent states.

For the universe to be a cellular robotic, all physical amounts must be separate. Besides, time and space must be separated. Although cellular robotics are capable of describing many basic physical phenomena, the quantum features of the universe are difficult to simulate using a conventional model such as cellular robotics.

The universe is quantum, and normal computers cannot simulate quantum systems, why? Because quantum mechanics is more exotic and counterintuitive compared to normal computers as for humans, in fact, to simulate a very small part of the universe, to transfer a few hundred atoms to one part of a second, a regular computer needs more memory space than the number of atoms in the entire universe, And to a longer time than the present age of the universe to end the simulation.

This is what led to the development of quantum computing models. Instead of relying on numbers - often numbers or bits - quantum computing relies on qubits. The difference between qubits and bits is that while bits can take one value from one of two values: 0 or 1, qubits can take a set of values ??that represent the superposition of states 0 and 1.

According to this principle, the universe is not a classic computer but rather a quantum computer, that is, it is a computer that does not process numbers but qubits. The quantum version of the universe is less radical than the traditional version, since the conventional version eliminates continuity from the universe, by claiming that deleting it allows classic computers to provide a literal description of the universe rather than an estimated one.

The universe is a physical system that is subject to computation; therefore, it can be simulated effectively using a quantum computer - the size of the universe itself - and because the universe supports quantum computing. It can be simulated using a quantum computer... It possesses a computing power that is no less than a quantum computer of the universe's size.

We have seen how physics laws can be used to perform quantum computing effectively, let us discover how a quantum computer can simulate physical laws.

Quantum simulation is the process by which a quantum computer simulates a quantum system. Because of the strange quantum properties, classic computers simulate quantum systems less effectively.

Still, because the quantum computer is itself a quantum system capable of highlighting all quantum properties, it can effectively simulate other quantum systems. Each part of the quantum system that you want to simulate is stored within a group that is stored within a group of The qubits are inside the quantum computer. The different interactions between these parts are transformed into logical quantum computational operations. The resulting simulations are so accurate that it is difficult to differentiate them from the simulated system.

No description of the universe was found as a computer before the twentieth century. The ancient Greeks atom indeed counted the universe as a form of interaction of small parts, but they did not explain whether these parts were information processing units.

Laplace invented a virtual object that calculates the future of the entire universe but considered it a separate entity from the universe and not the universe itself.

At the same time, Charles Babbage was not eager to use his device as a model for physical phenomena, unlike Alan Turing. The latter was interested in the origin of patterns and researched the topic.

The first explicit description of the universe as a vast computer was in 1956 in the science fiction novel "The Last Question" by Isaac Asimov. In this story, humans invent analog computers to help them explore their galaxy first and then other galaxies. The link between computing and physics began in the early 1960s by Rolf Landauer at IBM. The idea that the universe can be regarded as a computer was proposed by Fredkin and independently by Konrad Zuse. Frieden and Zeus suggested that the universe could be a type of classical computer called a cellular robot that contained rows of bits that interact with adjacent bits.

Stephen Wolfram recently developed and simplified this proposal. The idea of using cellular robotics as a basis for the theory of the universe seems interesting. Still, its problem is that classic computers are not able to reproduce quantum phenomena, such as quantum entanglement. Another reason, as mentioned earlier, is that simulating a small portion of the universe on a classic computer requires a volume equal to the size of the universe.

Therefore it is impossible to regard the universe as a classic computer as the cellular robot. In his research paper: Ultimate Physical Limits to Computation, Seth Lloyd demonstrated how the computing power of any physical system could be calculated by knowing the amount of energy present in the system and the size of the system, for example: Use these limits to calculate the maximum computing power of one kilogram of material stored in a liter volume one from space, the usual laptop weighs approximately one kilogram, and takes the size of one liter of space, so we will call the one-kilogram computer and one liter of super-laptop.

A super laptop is a 1-kg computer and 1 liter (regular laptop size), where every elementary particle was placed inside it for computing. A super laptop can perform 10 million logical calculations per second at ten thousand billion bits. What might be the power of a super laptop?

The first significant impediment to excellent computational performance is energy. The amount of energy determines the range of speed, for example: Let's take a single-bit electron, which moves here and there, whenever he has much energy whenever he moves quickly here and there and can change his bit-flip state quickly, the speed at which qubits change his condition is subject to a theory known as Margolus-Levitin.

The theory says that the maximum speed at which a specific physical system (for example, an electron) changes according to its energy. The more energy there is, the less time it takes for an electron to travel from one state to another, the theory is very general, as it does not care about any system that collects and processes information, but only concerned with the amount of energy that the physical system has to process this information.

To return to the supercomputer example, all the atoms and electrons that make up our computer have a temperature slightly higher than the room temperature. In every jolt related to temperature and not the state of the electron or atom, this is why the frequency at which the electrons change their states - from here to there or from 0 to 1 - with the frequency of the state of the atoms and electrons changes, and atoms change the state of their cube in the same frequency.

Everything Is a Computation

The Margulis-Levitin theory provides a relationship to calculate the value of the maximum velocity average in which a cube can change its state, we take the value of energy needed to change the state of the cube, multiplied by four and then divide it by Planck's constant so we get the number of times that the cube can change its state in The second one, it is a mathematical relationship that indicates that everything is a computation, when applying this relationship to the supercomputer, it was found that every atom and electron in a state of vibration in the computer changes its state about 30 trillion times per second. Of course, the rate at which electrons and atoms change their states is higher than the rate at which traditional computers change their states.

In his book, Seth Lloyd says the computer on which we write consumes energy to charge and discharge the capacitors whose homes keep millions of times the amount of energy that electrons and atoms consume to change their states. Still, it works 100,000 times slower than a supercomputer.

We can calculate the amount of energy needed to perform computing on a supercomputer using Einstein's famous phrase: the speed of light c and x, the mass of the computer m, the energy E where it represents, $E = mc^2$.

The supercomputer weighs 1 kg, and the speed of light is equal to (300 million meters per second). By applying the equation, we find that the supercomputer has 100 million billion (10^{17}) joules of energy for computing. With a more used unit, the computer has about 20 million millions kilos of energy. Another unit of energy calculation is the amount of energy released in a nuclear explosion, as the supercomputer has 20 million tons of TNT to perform computing operations, equivalent to the amount of energy released by a colossal hydrogen bomb.

When a supercomputer is at its highest potential, consuming all of its energy within it is something like a nuclear explosion. All of the primary particles that store and process information inside are vibrations below a billion-degree temperature. The supercomputer appears as a small part of the Big Bang, so the number of operations - per second - that our small and powerful computer can perform is enormous, a billion billion

billion billion billion billion processes per second. Suppose that all matter and energy in the universe are put to computing, how much will the resulting computer's power be? We can compute this cosmic computer's strength consisting of everything that makes up this universe in the same way that we calculated the power of the supercomputer at the top. As mentioned earlier: the amount of energy determines the speed.

The amount of energy in the universe, the large part of which is kept inside the atoms, has been calculated with a high degree of accuracy. If we compute all the number of atoms in stars and galaxies and add the matter in the interstellar cloud, the average density of the universe is equal to the amount of one hydrogen atom per cubic meter of space.

But there are other forms of energy in the universe. For example, the light also contains energy (but less than that in the atoms). The rate of rotation of the distant galaxies indicates the presence of distant, invisible sources of energy that take unknown forms.

The irregular frequency with which the universe extends indicates a form of form. Another form of energy is called: quintessence. However, the amount of energy present in these strange forms does not exceed ten times the strength of ordinary matter. This does not significantly affect the range of computational power of the universe. Before beginning to compute the universe's computing power to be clear about what we are calculating, the evidence gathered from visual observations indicates that the universe is unlimited and in continuous expansion in all directions.

Of course, in an infinite space, the energy is unlimited, so the number of operations performed and the number of bits in the universe is also endless. But the observations also tell us that the universe has an end age just under 14 billion years, and the information cannot travel faster than light because the universe has an ending age and because the speed of light is ending, the part of the universe from which we can obtain information about it is also finished. This part of the universe from which we can obtain information is contained within what is called the Cosmological horizon. Still, beyond the cosmic horizon, we can only guess what is going on there.

The number we will get represents the amount of computing possible within the cosmic horizon. Information processing operations outside the cosmic horizon do not affect any computational process that falls within the observed part of the universe since the Big Bang.

So when we say the calculation of the computing power of the universe, what is really meant is calculating the computing power of the part located within the horizon of the universe. The cosmic horizon continuously expands with time at a rate three times the speed of light. With the widening of the cosmic horizon, more things appear to us, and with it increases the amount of energy available to do computing.

The amount of computing that can be performed on the part of the horizon of the universe has increased over time since the Big Bang. The cosmic horizon is 42 billion light-years away from us. Each cubic meter of the portion within the horizon on the average contains the equivalent mass of a hydrogen atom per cubic meter, and every hydrogen atom releases energy.

By adding all the energy values in the universe, we find that it contains 100 million billion billion billion billion billion billion billion billion joules of energy. Most of this energy is free, that is, available for computing. To calculate the maximum speed at which the universe processes information, we apply the Margolis-Levitin theory: we take the amount of energy present in the cosmic horizon, multiplied by four and divide it by Planck's constant, and the result is: that every second a computer containing all the energy in the universe can perform 100,000 logical operations Arithmetic. In the 14 billion years since the creation of the universe, this virtual cosmic computer has delivered around 10,000 billion operations.

Now we know that the universe is computing, and we know the computing power of the universe. What now? What could be new to the idea of the universe as a giant quantum computer? We have prominent theories in quantum mechanics for elementary particles. What if these elementary particles process information and perform computing? Do we really need to think about developing a new theoretical framework for how the universe works? These questions are reasonable; the agreed-upon image of the universe from a physical perspective depends on a model that sees the universe as a machine. Contemporary physics is based on the mechanical model. Where does the ocean analyze based on the inner mechanisms? Indeed, this model is the basis of all modern science today.

Moreover, by referring to what Seth Lloyd says in his book and linking it with what he meant by Paolo Coelho with the spirit of the universe in his alchemical novel, we will conclude that the universe's sense is nothing but the common dimension of the universe. It is the computation, and according to Seth Lloyd, all the interactions between the particles in the universe are not transferred energy not only, but also information, in other words, molecules not only collide but also perform arithmetic operations with each other. The entire universe computes, each atom, electron, and elementary particle that stores a lot of information, and each time two particles collide, it processes these two slits.

Moreover, when you desperately want something, the entire universe meets to achieve it, because you are an integral part of this universe. You also transfer information to the particles in the universe, so the universe tries to accomplish everything you want. What is the link between us and the universe?

The answer is found by Nikola Tesla. Although physicist and electrician Nikola Tesla did not gain the reputation of Edison Thomas or Albert Einstein and did not reach the peak of richness and statues did not rise to him everywhere, he did not win the Nobel prize, but everyone who looks at the moon remembers Tesla nozzle. Everyone who studies physics knows very well who Tesla is and what he provided in electromagnetism. Everyone who holds a remote control must understand that Tesla is the first person who thought about making it.

He presented something valuable to physics, which is his research. He demonstrated the relationship between the electromagnetic field produced by a body and the total electrical currents that pass through it.

Through his research, we can prove that anything in the universe produces an electromagnetic field, it is sufficient to demonstrate that any object in the universe has an electric current passing through it, and if an electric current passes through it, we apply Tesla's research to demonstrate that any object produces an electromagnetic field.

Concerning the human being, to prove that he produces an electromagnetic field that is sufficient to return to the previous axis, we have learned that a person has an electric current in it, and accordingly we conclude according to Tesla's theorem that we also produce an electromagnetic field.

As for the other objects that belong to the universe, it is easy to prove the existence of an electric current that passes through it even if it is minimal, even if it is 0,00000001 amperes, all that concerns us is that it is not equal to zero, and to demonstrate that the electric current passes in everything in The universe suffices to remember that the electrons are the source of the current and that they are present in every atom in the universe. So everything in the universe passes through it.

You might think that the wood does not pass the current. Actually, it does, but in a tiny amount, it will not lit a lamp, so we consider it a dielectric or resistance to the current.

$Since everything in which the current in the universe passes, even in a small way, according to Tesla, everything in the universe produces an electromagnetic field. That field causes us to relate to the entire universe, which explains the transfer of information across the universe—the caller's voice over the phone from far to your room.

Thus, electromagnetic energy contributes to the realization of the law of attraction, but what are the problems that stand in the way of the fact that some things everyone wants quickly are not accomplished? The answer is that the forces of electromagnetic attraction are weak. To be weak, according to Tesla's theorem, the current must be low, and for the current to be low, resistance must be resisted.

So it is the cosmic determinants that settle how long something will happen, so you have to believe in everything that you desperately want until it is realized. All of this is only a logical and mathematical sequence.

Perhaps after all these indications that the universe is subject to computation and that everything in it is an account, we have not yet reached a comprehensive picture of the material universe.

So we will try to give a mathematical definition of everything in the material universe. Well, the thing that is common to all the findings is location.

Everything that belongs to the material universe can be identified with three elements: length, height, and width, i.e., X, Y, and Z, so every object in the universe is a function. Its starting level group (IR x IR) and its straight-line arrival group (IR) at the level to form a three-dimensional space, for example, the three-dimensional printer, prints the bottles of soft drinks similarly, so that the unique function of the bottle's shape is given in a mathematical form to transform it into reality as well, to transform it into reality.

Robots that make cars, parts of planes, or things that look perfect to you like phones rely on the same principle.

Well, everything in the material universe is a function, we will call it the function of the shape of each object, and let us count now, the physical universe is the sum of those functions, i.e., all these forms (mountain, planet, sea, tree, rock, etc.) as everyone knows the group of real functions is a real vector space (x x = F {F + F + F.... X times) where the sum of two functions is a function. Any real number of functions is also a function, as we know that then the vector space is also infinite, due to the existence of a family born to it that contains infinite elements.

We will now apply a linear application to the physical universe, i.e., to the set of those functions. This application is the function of motion (takes anything from the physical universe to the next temporal moment).

It is easy to prove that it is linear, it is sufficient to calculate the sum of two forms to obtain the sum of the two images, and also the image of a real number of forms is the same as the real number of the image of each form separately, if the number is negative, then this means that the bodies do not exist and therefore remains The application is linear in this case because the image of the void is the void, just as the number can be decimal. After all, the distance and the area are algebraic values (we mention this because a linear application such as a vector space is defined by fields such as IR).

This representation that we have given to time and movement is likewise found in the principle of motion in body mechanics.

Now that every application Linear can be linked to a matrice , then the motion application can also be linked to a matrix of a number, the columns and the lines in it are infinite, because the vector space that we define the application, which is the physical universe (the group of functions) has a dimension Infinite, that is, there is a family born with an infinite number of elements, it can be understood that the number of things in the universe cannot be counted. The matrix we will get changes from moment to moment because, at every moment, things in the universe move, and the positions and shapes of bodies change.

Since we cannot get that matrix because it is vast, we cannot know the future of things and how they move, so things are linked. It is possible to overlook a simple thing that we get to wrong calculations. In classical mechanics, predicting where the projectile occurs is roughly when external influences, such as wind forces and friction, are neglected. According to Laplace, the future of the universe itself can only be known and predicted by a higher entity that is separated from the universe, an entity that is able to get the entire matrix. and according to me just the Almighty God can obtain the matrix as a whole

Predicting the future or traveling to the past is difficult, experimentally, so how do we coexist with that? What is the secret behind changing the past and controlling the future? All of that will be dealt with in the next axis.

...THE PAST, PRESENT, AND THE FUTURE...

"I have an almost complete disregard of precedent, and a faith in the possibility of something better. It irritates me to be told how things have always been done. I defy the tyranny of precedent. I go for anything new that might improve the past."

Clara Barton

"The past is behind, learn from it. The future is ahead, prepare for it. The present is here, live it."

Thomas S. Monson

Everything Is a Computation

Humans are temporal beings, possess dates and tests these dates in stages, maintain a current record of their dates as they are revealed to them, and behave and head their eyes towards the future. Humans are created with an innate capacity to know all three-time dimensions (past, present, and future), but it is the nature of life experience and the different pressures that make it inclined to a temporal dimension and not others.

The concept of time casts a shadow over the movements of humans and their dwellings and interferes in the components of their personalities and throughout their lives, and perhaps the presence of humans in the world means their presence in time, it is presented to us as a movement emerging from the bowels of the past and passing the present on its way to the future. Time is a set of events with a beginning that delves into the deep past and extends through the present to the future. What makes time meaningful in our minds is the sequence of events between the past, present, and future.

The images that memory provides us with a series of lived events through the laws of retrieval make us aware of the relationships (before) and (after) that link them, as well as the capabilities of intuition, imagination, changes of aspiration, expectation, etc., which make a person look to their horizons Live it.

Our actions in this changing world always depend not only on the situation in which we find ourselves at that moment, but also on everything that we lived on and our future expectations, as the individual lives in the present, that is, their behavior is indicative of everything they set here and now, but their activities return them always to the past or to what will happen, the present has several dimensions, embodied by the statement of St. Augustine ("There is a present for past things and a present for present and present for future things," since the past, present and future as forces interacting with each other impose on the existence of the individual may be one dimension might prevail over another.

Time interconnection is the link between time zones. It is one of the basic concepts within the concept of time, where Heidegger believes that the activity of remembrance connects the past with the present, as it links the activity of expectation, intention, hope, and present to the future. Thus, in the spatial concept of the flow of time, it does not lose the past Absolutely - the memory restores it - and the future is not entirely unknown, as it acquires its substance from expectations and intentions. The individual can formulate representations of changes other than those he perceives in the present, and through these representations, his present can embrace the temporal perspectives of the past and the future that form the horizon of time. We build our past just as we build our future.

Adaptation appears to be the characteristic of this efficacy, as the individual must somehow free themselves from the state of change that they endure through their lives, by keeping the past available through memory and entering the future in advance by expectation. This control over time is an individual achievement conditional on all that determines personality such as age, environment, mood, experience, and each individual has a unique perspective since time is the succession of changes. However, each of these changes - excluding the present change - is present only as memory or expectation.

Suppose we review the writings of several authors. In that case, we find that there is disagreement about any time zone that bears the most significant importance for our understanding of the nature of time, and if we put these authors in categories according to the representation of the time zone that each of them thinks is the most important, we classify Bergson as believing "The domination of the past," because he insists that memory is the most crucial concept for understanding time, as for Whitehead and because he believes that all mental activities, remembering activities, sensations, and expectations exist in the present, we classify it as believing "in the dominance of the present," we classify Kelly as believing in the "hegemony of the future" because he emphasized the sustainable development of the personality and the individual's possession of the need to plan for the future.

On the subjective experience of time, time's objective and subjective characteristics have always been subjects of reflection and questioning. Therefore the philosophical and scientific formulations related to the nature of time are formed from broad and complex literature. In the years of the sixties and seventies of the last century, several studies were concerned with the relationship in time experience. Between the characteristics of the personality and its dynamism, these studies have supported the idea that the distinct ways of passing through the experience of time and its use vary significantly between individuals and on dimensions that can be estimated and measured and that these differences are meaningfully related to the characteristics of the personality.

When time is inherently seen as an objective reality that exists independently of human consciousness, elite philosophers such as Eliade, Kummel, physicists, Cunningham, and psychologists Piaget and Wallace, & Rabin saw time as a subjective experience of movement during the void.

The division of time into three sections is a cross-cultural phenomenon, as a person possesses memories about what had gone before and calls it past, they pass through the experience of the current moment and calls it present, and expect what has not yet to come and name it future, and connects these dimensions in a unit called time. There is a difference in establishing the internal link between these areas.

The process of remembering is only possible when the intermediate mechanisms in the brain of the organism can visualize and create things and events that are not parts of their immediate sensory environment. Therefore it is challenging for an adult person to fully define himself in the present for an extended period of time.

The interconnected events of the past and the future do not dissolve the intersection of consciousness, integrated into its concurrent patterns of action and meaning, and this process is related to a great extent to the emergence of self-awareness, as the person gradually represents their changing memories of their past, to a sense of identity and the continuity of self, the expectations of the future become an integral part of the meaning of the present experience.

The future is mainly created from the contents of past experience, from the sense of continuity and from the change that can be expected regularly from this continuity and vice versa, when changing circumstances in the present create new concepts for the future, then the reconstruction of the past will change at most accordingly.

One way to look at time correlation is to see it as an individual's tendency to integrate the past and the future into the present region, and this is what therapists who seek to help patients may aim to be "here and now" or "present" and the idea appears consistent, as the individuals who see The regions of the past and the future as combined with the present will tend to evaluate the present more positively. These results should be seen alongside the Cottle results saying that individuals with a high temporal correlation are less anxious and smarter than individuals with a low temporal correlation.

In some people, the regions of the past, present, and future are not interconnected, nothing is united by sequence, and this perception is called: atomic, while other people see the time zones linearly, moments that follow moments, and each of them, and once we test them, they disappear In a way that cannot be restored, in the past, and correspondingly, every experience has its infinite moment, called this continuous realization. There are other individuals whose individual moments seem to extend, as the areas of the past, present, and future appear to overlap, and the child does not have a clear concept of time, but only temporal impressions (wait, for example). Under the influence of neurophysiological maturity and the achievement of logical continuity, events occur in order to form a time sequence system, and time appears irreversible and organized in the past, present, and future. The concept of permanence is formed and becomes measurable.

Furthermore, suppose time is a stage marked by a succession of events, adjustment of state, and change. In that case, time is an intellectual structure that seeks sovereignty over the transit, they are an essential aspect of our sensory experience: change, as defined by William Leibens: "The system of successive phenomena, and there would be no time if the world were still and steady, even though the day did not follow the night and the spring followed the winter and the pleasure followed the sadness." This is consistent with Albert Einstein's theory of time, as it is also related to change and speed.

Human time has two aspects: a physical and social aspect based on the succession of night and day, explained by the hour and diary, and a personal subjective aspect that varies from individual to individual and from moment to moment, and the pension time that fluctuates with our psychological states, our interests, and our actions is not a homogeneous dimension; Sometimes it seems that it is expanding slowly, and at other times it shrinks and moves quickly, sickness, unemployment, and dimension-free time from its essence, so it seems to us in its cases that it elongates indefinitely. As for the dense, arduous, and enjoyable activity, on the contrary, it gives it an intensity that makes it seem to us very short (at least at the present moment, because it is the stages rich in strong impressions that seem to us in terms of the most extended past).

Moreover, if the time is taking place in aging faster than its flow in childhood and adolescence, the reason is that aging differs profoundly from these two stages of a person's life. Childhood and adolescence are a time of acquisitions, always new impressions, and a struggle for the individual to secure himself a situation, the discovery of love, etc. As for aging, on the contrary, it corresponds to a stage where almost everything has happened before, and where nothing is surprising or wonder, and where time is empty because it takes place in a world whose activities are diminished, monotonous, and devoid of great interest. The concept of time falls within a general continuity, the continuity of adapting to the world that surrounds us, the awareness of a period of time and the rhythm of action and rhythm in the phenomena that successive and occur, makes us able to anticipate the phases of change and make our behaviors compatible with them.

Reaching this level of the book only indicates that you have begun to realize the computation which we are subject to and that you have become see things in a logical way, and began to believe that everything is a logical, equivalence sequence linked to the computation. I want to play a game of logic with you, I know that you will focus logically and intensely, this game we will pave the way for the fundamental wisdom that we will come out of this axis.

In this game, I will ask you a few questions, and you try to answer me. The first question we begin with is: Where is the letter that you read at the last moment? Where are all those words that you have read now? Where is this word that you read now? Focus on each letter like a moment. Whenever you read a letter, the moment associated with it will go. Let me ask you again. Where is that moment? You do not know where it is, but all you know and feel is the next moment, also this moment, and the same thing goes for this moment too, which also went with the previous moment.

Well, do not bother thinking about the last game; you did not even know what the future of the game would be like when we played it. That future went with the previous moments. What you experience now is the next moments, which were a future when I asked you where was the future that has now been with all those moments, so where are they? The moment that you feel is only the next moment, meaning all you feel is the present. Game over.

All that can be drawn from that game is that the past exists only in our mind and that the future also becomes from the past when it's already happened, and all that connects them is the moment we feel which is the present moment, the moment that you read right now. What you do at the moment is what turns into a good or bad future, and that future turns into a sad or happy past. Depending on the computation you made at a particular moment, it might have a significant impact on one's entire future.

Some books have previously talked about this effect, such as the book Pistachio Theory, and there is also a theory on this subject called the theory of butterfly effect. To understand how this interconnection occurs between moments between them, we will talk about how a small decision or an act can have countless possibilities.

Imagine that you are leaving your house on a sunny day, carrying a bottle of water. After drinking all of the water, you threw that bottle on the road.

A week later, that bottle got affected because it was run over by a car. Simultaneously, due to the steam of a jug located different city, it has rained heavily in your city, so the heavy water carried the bottle to the passage of rainwater, causing the passage to be closed, which led to a flood, then when you went out to buy something from outside and you passed by the curve which had a lot of water in due to the closed passage, a car passed quickly and made some water fly from the ground to your clothes, and it got dirty. You did not bear that, and you had an intense fight with the car owner because of that until you killed him, then the police came.

A lawyer is waiting for his wages because of a small bottle of water. You have set a bad future for yourself because you do not throw the bottle in the place designated for the garbage, and that future will also turn into a sad past, all because of an event at that given moment.

There are many mathematical interpretations of the butterfly effect theory. For example, in mathematics it is called série harmonica, it has no end when n goes to infinity, although n / 1 turns to zero when n translates to infinity, we conclude from this example that a small number can become meaningful over time. There are many films that talk about such probabilistic patterns, and if you want to understand the butterfly effect theory well, I advise you to watch the MR NOBODY movie.

Everything Is a Computation

Everything that happens to you is computational, whether good or bad, so I want to share with you the wisdom that we have come from this axis, if you apply it you will get a glorious future that will turn into a memorable past.

"... the recent past in the distant past was the future, and the near future will become past in the distant future, and the calculations you make now will determine your future, so take advantage of the moment ..."

... *DEATH* ...

"No one wants to die. Even people who want to go to heaven don't want to die to get there. And yet death is the destination we all share. No one has ever escaped it. And that is as it should be, because Death is very likely the single best invention of Life. It is Life's change agent. It clears out the old to make way for the new."

Steve Jobs

Everything Is a Computation

We have previously identified human comprehensively and accurately. The most important fact we came to know is that human is nothing but a set of calculations, which can be represented by an automatic dimension, hence, programmatic. All of that that we identified in the human axis indicates that the death will be the end of a mathematical and computational stage as well as the passage to another stage of perception where all human information and memories will be converted into energy. To understand that, imagine that there is a person who documented all the events of his or her entire life, and uploaded it on a website, and then broke his phone, where will all the uploaded data be transformed? They are in space, and they have become energy.

Whatever exists after death is not subject to the same laws that our world is subject to. In this world, you can say: This mountain is huge than this mountain, but after death, there are no such comparisons. The afterlife is beyond time and place, so what we are aware of here, the questions we ask here, the statements we make here, etc., will be illogical and have no meaning whatsoever. When we die, all we take with us are our memories, and all that we have done during our lives, just as we Muslims and Christians and Jews also believe in what we call the resurrection or the afterlife, that is not far from science and logic.

The theoretical researcher Dr. Robert Lanza argues that a person dies only physically, while the mind transforms into potential energy that is released from the body to the outside in a process that the scientist called "vital central." Moreover, according to what the "Huffington Post" quoted from the British newspaper "The Guardian," the theory holds that there is no such thing as death. Therefore, death is according to the theory of the scientist and theoretical researcher Dr. Robert Lanza is the death of the body only, but the mind is energy within the body that is released outside when the life of the physical being stops in a process called "Biocentrism."

The theory of the scientist Lanza is essentially based on expansion and detail in a famous saying by the scientist Einstein, in which the physicist said: "Energy cannot be created or destroyed, it can only be changed from one form to another."

Thus, when our bodies die, the energy of our perception may continue. However, at the level of the quantum dimension, therefore Lanza imagines that our perception continues to exist but in a parallel universe, and Lanza refers to the principle of uncertainty or skepticism which is a theory that came out in 1927 launched by German physicist Werner Heisenberg; The theory says that the speed and location of an object can be measured at the same time, somewhere where molecules vault. "If there really is a world somewhere where molecules converge in their existence, this means that we must be able to measure all their properties, but we cannot," says Lanza.

Take, for instance, the double-slit experiment; when we see molecules without offspring (smaller than an atom) or light that passes through cracks in a barrier. With hard body strikes, they are like small bullets fired through one or another hole. However, if scientists do not "observe" the path of the molecule, then its behavior looks like waves that allow it to cross through both slits simultaneously," Lanza also says, "Why does our observational perspective change what is going on? Answer: Because the reality is a computational process that requires our awareness of it.

All in all, Dr. says. Lanza that there is nothing without perception and that what we see is nothing but a mere perspective, in other words, he says: All that we see is nothing but information within our perception that is stored in the body that will destroy itself sooner or later, in addition to that time and space are tools in our hands to understand the role and location of every piece in the mystery of the universe and that there is no death in a world without time and space.

Everything Is a Computation

What we draw from all of this is that we must do good in our lives, and prepare life as a painting, and draw beautiful memories on it, perhaps it may remain with us forever, so a person must have morals and human values, for example, an individual who killed a person and then committed suicide, they will remain in no place or time regretting what they did and will remain in the idea of why I killed that person and killed myself, they will, therefore, remain in nothingness. Nothing is close to describing it as hell, where a person burns to the death then gets resurrected to be burnt to death, and so on, it becomes empty. Who could be passive (dead) and positive (alive) at the same time? It is the zero, indeed, the zero. As for those who did good in life and contributed to the improvement of humanity, their names will remain immortal, and they will live outside time and place eternally happy.

...GOD AND COMPUTATION...

"That deep emotional conviction of the presence of a superior reasoning power, which is revealed in the incomprehensible universe, forms my idea of God.."

Albert Einstein

Everything Is a Computation

Everything Is a Computation

The book is near to be completed, and things are getting serious. Reaching this level of the book should have significance. The letters, words, ideas, assets, creatures, the universe, time, and death have combined in one word, which is the computation. Everything has become a computation, a logical and mathematical sequence, so who is behind this computation? Who set it up?

Perhaps we must travel to the distant past at the beginning of the universe to answer these very questions, and this is what nobody hasn't been able to do. Who among us can travel to the past?

Can we answer all of these questions without traveling to the past? Or will the answer be only from the present? As we have already known in the past, present, and future axis, the present is linked to the past, meaning everything that I did in the past is what extends to the present.

Also, everything that I did in the present affects the future; this link means that the beginning, the current moment, and the end are intrinsically linked, meaning that we can understand what happened in the past with the computation in the present, and we can also understand the future with what is happening in the present.

Many theories could explain the methods of the beginning of time or the beginning of the universe from events in the present. Among these theories, the Big Bang theory, one of whose greatest pioneers was Stephen Hawking, this theory, which is based on the idea that the universe expanded from a size smaller than an electron to nearly its current size within a tiny fraction of a second after which all galaxies formed and Planets, moons, and stars.

This Big Bang explosion is nothing but a chemical-physical equation that has a mathematical dimension, meaning that the explosion also is a calculation, so if it had not been a calculation, many other calculations would not have appeared, including the law of soil and the law of stone and the human mind, etc. Moreover, since behind every calculation there is a solid logic, then there must be a logic behind the Big Bang, and this logic leads us to the logic of the existence of a God who created the universe and gave the beginning to it, it does not matter how that happened. All that matters is that there is a God behind all these calculations. I do not want to describe God, nor discuss the way of worshiping God. While I was writing the book, I was trying to get as far as possible away from a bias for any religion, because I want to explain everything using computation and keep my faith and religious philosophy to myself. However, I am against the idea that there is no creator of the universe, because it contradicts the idea of the book, that everything is a logical arithmetic sequence, and the sequence for it to be complete there must be a higher being (God) who is not subject to the laws to which we are subject and possesses the entire matrix (universe).

Along the way to spread this idea, I was always meeting with people who were against it, where they argued that the universe was random and coincidental, just as I met with people who agreed with my ideas, and they believed that the universe has a creative designer.

Perhaps through what we previously discussed, there are two theories that attempt to understand the universe. The interpretation of the first universe is chaotic and based on randomness and chance, and the second depends on order, accuracy, and creativity. From my perspective, I see that creativity theory (intelligent design) is the closest to reality, there is no randomness and chance in a universe governed by laws and the numbers and the calculation, not even a small protein can be created by chance. What about these many living organisms that contain trillions of proteins? How can they exist by chance with such precision and beauty? Look at the fish how it formed.

Doesn't it make sense for a protein to be created by chance, of course not, and that was not what Charles Eugene Jay, the Swiss naturalist, could prove by calculating the probability of creation by chance for one protein molecule, the latter known to consist of 4 different elements. However, he assumed that the protein molecule consisted of only two elements out of the total of 2000 atoms that compose it after this simplification, it has been found that the probability of creating a protein by chance is (approximately zero, $1/10^{160}$), so if we take this result into account within the age and size of our planet, then creating such This molecule takes 1,0243 billion years under conditions of 51,014 vibrations per second. Accordingly, there is no possibility that life may have arisen by chance during the 4.5 billion years that are assumed to be Earth's age.

This calculation was repeated by Manfred Eigen of the Max Planck Institute for Biochemistry in Göttingen, Germany, who won the Nobel Prize in Chemistry in 1968, and he proved that all the water on our planet is not enough to randomly produce a single protein molecule even if the entire universe is full of chemicals that unite always with each other, if the age of the universe is not sufficient to regulate one protein by accident, it is not wise to assume that randomness is able to regulate the universe, and did you know that in order for a person to form, 250 proteins must be gathered and the age of the universe is not sufficient to create one by chance, and then we get the first living cell and then we need to retry and succeed 37.2 trillion times, and to arrange again by chance, so you have the first human being, and then you need to create thousands of systems; In order for this organism to survive, including the compound of water and air, gravitational control, weak and strong nuclear force, and electromagnetic energy, and the sun and moon and the conditions necessary for the life and growth of the organism need, it is impossible in particular, and this scientific evidence contradicts chance and nullifies the idea that there is no creator of the universe.

Some may see that chance and randomness have evolved with time, but what these people do not know is that time itself is based on a precise calculation and logic. All that has happened to the cell in terms of evolution according to the theory of evolution are not random mutations, but rather arithmetic operations and mutation are only accounts that occur via a change in genetic information and Biological, genetic information encoded in deoxyribonucleic acid sequences, and chromosomes contained in DNA, or in DNA sequences in the case of some viruses. DNA is like a double chain, and the parts that make up this chain are nucleotides (also called nitrogenous bases).

The mutation can make changes to DNA or RNA strings in different ways. They may alter the nucleotide sequence arrangement or number by inserting one or more bases or deleting one or more nitrogenous bases or a hopping gene. The important thing is that all of this is merely programmed calculations since the origins of the universe, and, logically, there be a programmer for it and a creator of it.

Several religions talk about that. I do not want to to take sides here, as I mentioned earlier. I will leave that for the future. Maybe I will write a book about religions from my computational point of view. Now all that matters is that everything is a computation, and behind all that is God who created us and designed this universe.

... *CONCLUSION* ...

"Everything we do, every thought we've ever had, is produced by the human brain. But exactly how it operates remains one of the biggest unsolved mysteries, and it seems the more we probe its secrets, the more surprises we find."

Neil deGrasse Tyson

There is a Creator of the universe who set all these complicated calculations in it. As soon as we realize it, we will have reached the significance of our existence, which is to respect that Creator. I hope that I was able to prove that.

There is a story about a doctor called Burhev who employed some criminals in scientific experiments and research in exchange for financial compensation for their families, and that their names be written in the history of scientific research, and in coordination with the Supreme Court and in the presence of a group of scientists interested in his experiences. Dr. Burhev made a man who sentenced to death to sit on a chair, and it was agreed that the sentenced man would be executed by clearing his blood under the pretext of studying the changes that the body undergoes during that state.

Dr. Burhev blindfolded the man's eyes, then installed two thin hoses on his body, starting from his heart and ending at his elbows, and pumped into them warm water at the same temperature as the body drips at his elbows, and placed two buckets under his hands and at a suitable distance so that water drips from the hoses into them so that it will sound like blood dripping as if the man's blood is coming out and dripping in the buckets.

After a few minutes, the researchers noticed pale and yellowishness in the entire body of the man sentenced to death, they stood close to examine the man, and when they exposed his face, a few minutes of suffering the man died! He died because he believed that he was bleeding to death, while actually, he did not lose a drop of blood! Moreover, the worst thing is that he died at the same time that it took blood to shed from the body and cause death! This means that the mind gives orders to all parts of the body to stop working in response to the perfect imagination exactly as it responds to the truth!! Therefore, pay close attention to the negative thoughts that are pumped into your mind, whether by yourself or by other people who want to discourage you.

Whether positive or negative, the arithmetic brain messages determine the approach of our life that we live in. Through all of the above, we conclude that everything is related to a logical arithmetic sequence, so we must deal with this matter seriously and change the way we think ourselves, and look for a goal to live and fight for, whatever happens to us, whether good or bad, is a computational message that we are the ones who carried it out... The end.

To be continued

SOME SCIENTIFIC CONCEPTS AND THEORIES MENTIONED IN THE BOOK:

Vector space

A vector space (also called a linear space) is a collection of objects called vectors, which may be added together and multiplied ("scaled") by numbers, called scalars. Scalars are often taken to be real numbers, but there are also vector spaces with scalar multiplication by complex numbers, rational numbers, or generally any field. The operations of vector addition and scalar multiplication must satisfy certain requirements, called axioms, listed below, in § Definition. For specifying that the scalars are real or complex numbers, the terms real vector space and complex vector space are often used.

Euclidean vectors are an example of a vector space. They represent physical quantities such as forces: any two forces (of the same type) can be added to yield a third, and the multiplication of a force vector by a real multiplier is another force vector. In the same vein, but in a more geometric sense, vectors representing displacements in the plane or in three-dimensional space also form vector spaces. Vectors in vector spaces do not necessarily have to be arrow-like objects as they appear in the mentioned examples: vectors are regarded as abstract mathematical objects with particular properties, which in some cases can be visualized as arrows.

Vector spaces are the subject of linear algebra and are well characterized by their dimension, which, roughly speaking, specifies the number of independent directions in the space. Infinite-dimensional vector spaces arise naturally in mathematical analysis, as function spaces, whose vectors are functions. These vector spaces are generally endowed with additional structure, which may be a topology, allowing the consideration of issues of proximity and continuity.

Among these topologies, those that are defined by a norm or inner product are more commonly used, as having a notion of distance between two vectors. This is particularly the case of Banach spaces and Hilbert spaces, which are fundamental in mathematical analysis.

Historically, the first ideas leading to vector spaces can be traced back as far as the 17th century's analytic geometry, matrices, systems of linear equations, and Euclidean vectors. The modern, more abstract treatment, first formulated by Giuseppe Peano in 1888, encompasses more general objects than Euclidean space, but much of the theory can be seen as an extension of classical geometric ideas like lines, planes and their higher-dimensional analogs.

Today, vector spaces are applied throughout mathematics, science and engineering.

They are the appropriate linear-algebraic notion to deal with systems of linear equations. They offer a framework for Fourier expansion, which is employed in image compression routines, and they provide an environment that can be used for solution techniques for partial differential equations. Furthermore, vector spaces furnish an abstract, coordinate-free way of dealing with geometrical and physical objects such as tensors.

This in turn allows the examination of local properties of manifolds by linearization techniques. Vector spaces may be generalized in several ways, leading to more advanced notions in geometry and abstract algebra.

Electromagnetism

Electromagnetism is a branch of physics involving the study of the electromagnetic force, a type of physical interaction that occurs between electrically charged particles. The electromagnetic force is carried by electromagnetic fields composed of electric fields and magnetic fields, and it is responsible for electromagnetic radiation such as light. It is one of the four fundamental interactions (commonly called forces) in nature, together with the strong interaction, the weak interaction, and gravitation.

At high energy the weak force and electromagnetic force are unified as a single electroweak force.

Lightning is an electrostatic discharge that travels between two charged regions.

Electromagnetic phenomena are defined in terms of the electromagnetic force, sometimes called the Lorentz force, which includes both electricity and magnetism as different manifestations of the same phenomenon.

The electromagnetic force plays a major role in determining the internal properties of most objects encountered in daily life.

The electromagnetic attraction between atomic nuclei and their orbital electrons holds atoms together. Electromagnetic forces are responsible for the chemical bonds between atoms which create molecules, and intermolecular forces. The electromagnetic force governs all chemical processes, which arise from interactions between the electrons of neighboring atoms.

There are numerous mathematical descriptions of the electromagnetic field. In classical electrodynamics, electric fields are described as electric potential and electric current. In Faraday's law, magnetic fields are associated with electromagnetic induction and magnetism, and Maxwell's equations describe how electric and magnetic fields are generated and altered by each other and by charges and currents.

The theoretical implications of electromagnetism, particularly the establishment of the speed of light based on properties of the "medium" of propagation (permeability and permittivity), led to the development of special relativity by Albert Einstein in 1905.

Capacitor

A capacitor is a device that stores electrical energy in an electric field. It is a passive electronic component with two terminals.

The effect of a capacitor is known as capacitance. While some capacitance exists between any two electrical conductors in proximity in a circuit, a capacitor is a component designed to add capacitance to a circuit. The capacitor was originally known as a condenser or condensator. This name and its cognates are still widely used in many languages, but rarely in English, one notable exception being condenser microphones, also called capacitor microphones.

The physical form and construction of practical capacitors vary widely and many types of capacitor are in common use. Most capacitors contain at least two electrical conductors often in the form of metallic plates or surfaces separated by a dielectric medium. A conductor may be a foil, thin film, sintered bead of metal, or an electrolyte. The nonconducting dielectric acts to increase the capacitor's charge capacity. Materials commonly used as dielectrics include glass, ceramic, plastic film, paper, mica, air, and oxide layers. Capacitors are widely used as parts of electrical circuits in many common electrical devices. Unlike a resistor, an ideal capacitor does not dissipate energy, although real-life capacitors do dissipate a small amount. (See Non-ideal behavior) When an electric potential, a voltage, is applied across the terminals of a capacitor, for example when a capacitor is connected across a battery, an electric field develops across the dielectric, causing a net positive charge to collect on one plate and net negative charge to collect on the other plate. No current actually flows through the dielectric. However, there is a flow of charge through the source circuit. If the condition is maintained sufficiently long, the current through the source circuit ceases. If a time-varying voltage is applied across the leads of the capacitor, the source experiences an ongoing current due to the charging and discharging cycles of the capacitor.

The earliest forms of capacitors were created in the 1740s, when European experimenters discovered that electric charge could be stored in water-filled glass jars that came to be known as Leyden jars. In 1748, Benjamin Franklin connected a series of jars together to create what he called an "electrical battery", from their visual similarity to a battery of cannon, which became the standard English term electric battery. Today, capacitors are widely used in electronic circuits for blocking direct current while allowing alternating current to pass. In analog filter networks, they smooth the output of power supplies. In resonant circuits they tune radios to particular frequencies. In electric power transmission systems, they stabilize voltage and power flow.[2] The property of energy storage in capacitors was exploited as dynamic memory in early digital computers,[3] and still is in modern DRAM.

Boolean algebra

In mathematics and mathematical logic, Boolean algebra is the branch of algebra in which the values of the variables are the truth values true and false, usually denoted 1 and 0 respectively. Instead of elementary algebra where the values of the variables are numbers, and the prime operations are addition and multiplication, the main operations of Boolean algebra are the conjunction (and) denoted as ∧, the disjunction (or) denoted as ∨, and the negation (not) denoted as ¬. It is thus a formalism for describing logical operations in the same way that elementary algebra describes numerical operations.

Boolean algebra was introduced by George Boole in his first book The Mathematical Analysis of Logic (1847), and set forth more fully in his An Investigation of the Laws of Thought (1854). According to Huntington, the term "Boolean algebra" was first suggested by Sheffer in 1913,although Charles Sanders Peirce in 1880 gave the title "A Boolean Algebra with One Constant" to the first chapter of his "The Simplest Mathematics". Boolean algebra has been fundamental in the development of digital electronics, and is provided for in all modern programming languages. It is also used in set theory and statistics.

Turing machine

A Turing machine is a mathematical model of computation that defines an abstract machine, which manipulates symbols on a strip of tape according to a table of rules.Despite the model's simplicity, given any computer algorithm, a Turing machine capable of simulating that algorithm's logic can be constructed.

The machine operates on an infinite memory tape divided into discrete "cells".The machine positions its "head" over a cell and "reads" or "scans" the symbol there. Then, as per the symbol and the machine's own present state in a "finite table" of user-specified instructions, the machine (i) writes a symbol (e.g., a digit or a letter from a finite alphabet) in the cell (some models allow symbol erasure or no writing), then (ii) either moves the tape one cell left or right (some models allow no motion, some models move the head),then (iii) (as determined by the observed symbol and the machine's own state in the table) either proceeds to a subsequent instruction or halts the computation.

The Turing machine was invented in 1936 by Alan Turing, who called it an "a-machine" (automatic machine). With this model, Turing was able to answer two questions in the negative: (1) does a machine exist that can determine whether any arbitrary machine on its tape is "circular" (e.g., freezes, or fails to continue its computational task); similarly, (2) does a machine exist that can determine whether any arbitrary machine on its tape ever prints a given symbol. Thus by providing a mathematical description of a very simple device capable of arbitrary computations, he was able to prove properties of computation in general—and in particular, the uncomputability of the Entscheidungsproblem ('decision problem').

Turing machines proved the existence of fundamental limitations on the power of mechanical computation. While they can express arbitrary computations, their minimalist design makes them unsuitable for computation in practice: real-world computers are based on different designs that, unlike Turing machines, use random-access memory.

Turing completeness is the ability for a system of instructions to simulate a Turing machine. A programming language that is Turing complete is theoretically capable of expressing all tasks accomplishable by computers; nearly all programming languages are Turing complete if the limitations of finite memory are ignored.

A Turing machine is a general example of a central processing unit (CPU) that controls all data manipulation done by a computer, with the canonical machine using sequential memory to store data. More specifically, it is a machine (automaton) capable of enumerating some arbitrary subset of valid strings of an alphabet; these strings are part of a recursively enumerable set. A Turing machine has a tape of infinite length on which it can perform read and write operations.

Assuming a black box, the Turing machine cannot know whether it will eventually enumerate any one specific string of the subset with a given program. This is due to the fact that the halting problem is unsolvable, which has major implications for the theoretical limits of computing.

The Turing machine is capable of processing an unrestricted grammar, which further implies that it is capable of robustly evaluating first-order logic in an infinite number of ways. This is famously demonstrated through lambda calculus.

A Turing machine that is able to simulate any other Turing machine is called a universal Turing machine (UTM, or simply a universal machine). A more mathematically oriented definition with a similar "universal" nature was introduced by Alonzo Church, whose work on lambda calculus intertwined with Turing's in a formal theory of computation known as the Church–Turing thesis. The thesis states that Turing machines indeed capture the informal notion of effective methods in logic and mathematics, and provide a precise definition of an algorithm or "mechanical procedure". Studying their abstract properties yields many insights into computer science and complexity theory.

Spacetime

In physics, Spacetime is any mathematical model which fuses the three dimensions of space and the one dimension of time into a single four-dimensional manifold. Spacetime diagrams can be used to visualize relativistic effects, such as why different observers perceive where and when events occur differently.

Until the 20th century, it was assumed that the 3-dimensional geometry of the universe (its spatial expression in terms of coordinates, distances, and directions) was independent of one-dimensional time. However, in 1905, Albert Einstein based a work on special relativity on two postulates:

The laws of physics are invariant (i.e., identical) in all inertial systems (i.e., non-accelerating frames of reference)
The speed of light in a vacuum is the same for all observers, regardless of the motion of the light source.

The logical consequence of taking these postulates together is the inseparable joining together of the four dimensions—hitherto assumed as independent—of space and time. Many counterintuitive consequences emerge: in addition to being independent of the motion of the light source, the speed of light is of constant magnitude regardless of the frame of reference in which it is measured; the distances and even temporal ordering of pairs of events change when measured in different inertial frames of reference (this is the relativity of simultaneity); and the linear additivity of velocities no longer holds true.

Einstein framed his theory in terms of kinematics (the study of moving bodies). His theory was an advance over Lorentz's 1904 theory of electromagnetic phenomena and Poincaré's electrodynamic theory. Although these theories included equations identical to those that Einstein introduced (i.e., the Lorentz transformation), they were essentially ad hoc models proposed to explain the results of various experiments—including the famous Michelson–Morley interferometer experiment—that were extremely difficult to fit into existing paradigms.

In 1908, Hermann Minkowski—once one of the math professors of a young Einstein in Zürich—presented a geometric interpretation of special relativity that fused time and the three spatial dimensions of space into a single four-dimensional continuum now known as Minkowski space. A key feature of this interpretation is the formal definition of the spacetime interval. Although measurements of distance and time between events differ for measurements made in different reference frames, the spacetime interval is independent of the inertial frame of reference in which they are recorded.

Minkowski's geometric interpretation of relativity was to prove vital to Einstein's development of his 1915 general theory of relativity, wherein he showed how mass and energy curve flat spacetime into a pseudo-Riemannian manifold.

Quantum mechanics

Quantum mechanics (QM; also known as quantum physics, quantum theory, the wave mechanical model and matrix mechanics), part of quantum field theory, is a fundamental theory in physics. It describes physical properties of nature on an atomic scale.

Classical physics, the description of physics that existed before the theory of relativity and quantum mechanics, describes many aspects of nature at an ordinary (macroscopic) scale, while quantum mechanics explains the aspects of nature at small (atomic and subatomic) scales, for which classical mechanics is insufficient.

Most theories in classical physics can be derived from quantum mechanics as an approximation valid at large (macroscopic) scale. Quantum mechanics differs from classical physics in that energy, momentum, angular momentum, and other quantities of a bound system are restricted to discrete values (quantization), objects have characteristics of both particles and waves (wave-particle duality), and there are limits to how accurately the value of a physical quantity can be predicted prior to its measurement, given a complete set of initial conditions (the uncertainty principle).

Quantum mechanics arose gradually, from theories to explain observations which could not be reconciled with classical physics, such as Max Planck's solution in 1900 to the black-body radiation problem, and the correspondence between energy and frequency in Albert Einstein's 1905 paper which explained the photoelectric effect.

Early quantum theory was profoundly re-conceived in the mid-1920s by Neils Bohr, Erwin Schrödinger, Werner Heisenberg, Max Born and others. The first put together interpretation of quantum mechanics is the Copenhagen Interpretation, this interpretation was developed by Neils Bohr and Werner Heisenberg in Copenhagen during the 1920s. The modern theory is formulated in various specially developed mathematical formalisms. In one of them, a mathematical function, the wave function, provides information about the probability amplitude of energy, momentum, and other physical properties of a particle.

Theory *of* computation

In theoretical computer science and mathematics, the theory of computation is the branch that deals with how efficiently problems can be solved on a model of computation, using an algorithm. The field is divided into three major branches: automata theory and formal languages, computability theory, and computational complexity theory, which are linked by the question: "What are the fundamental capabilities and limitations of computers?".

In order to perform a rigorous study of computation, computer scientists work with a mathematical abstraction of computers called a model of computation.

There are several models in use, but the most commonly examined is the Turing machine.

Computer scientists study the Turing machine because it is simple to formulate, can be analyzed and used to prove results, and because it represents what many consider the most powerful possible "reasonable" model of computation (see Church–Turing thesis). It might seem that the potentially infinite memory capacity is an unrealizable attribute, but any decidable problem solved by a Turing machine will always require only a finite amount of memory.

So in principle, any problem that can be solved (decided) by a Turing machine can be solved by a computer that has a finite amount of memory.

Biocentrism

In 2007 Lanza's article "A New Theory of the Universe" appeared in The American Scholar. The essay addressed Lanza's idea of a biocentric universe, which places biology above the other sciences.Lanza's book Biocentrism: How Life and Consciousness are the Keys to Understanding the Universe followed in 2009, co-written with Bob Berman.

In 2016 a second book, Beyond Biocentrism: Rethinking Time, Space, Consciousness, and the Illusion of Death was published.

Lanza's biocentric hypothesis met with a mixed reception. Deepak Chopra called "Lanza's insights into the nature of consciousness original and exciting" and stated that "his theory of biocentrism is consistent with the most ancient wisdom traditions of the world which says that consciousness conceives, governs, and becomes a physical world. It is the ground of our Being in which both subjective and objective reality come into existence."

David Thompson, an astrophysicist at NASA's Goddard Space Flight Center said that Lanza's "work is a wake-up call". Nobel laureate (Physiology or Medicine) E. Donnall Thomas stated that "Any short statement does not do justice to such a scholarly work. The work is a scholarly consideration of science and philosophy that brings biology into the central role in unifying the whole."

Arizona State University physicist Lawrence Krauss stated: "It may represent interesting philosophy, but it doesn't look, at first glance, as if it will change anything about science."

Wake Forest University scientist Anthony Atala stated "This new theory is certain to revolutionize our concepts of the laws of nature for centuries to come."[40] In USAToday Online, astrophysicist and science writer David Lindley asserted that Lanza's concept was a "...vague, inarticulate metaphor..." and stated that "...I certainly don't see how thinking his way would lead you into any new sort of scientific or philosophical insight. That's all very nice, I would say to Lanza, but now what?" Daniel Dennett, a Tufts University philosopher, said he did not think the concept meets the standard of a philosophical theory. "It looks like an opposite of a theory, because he doesn't explain how it [consciousness] happens at all. He's stopping where the fun begins."

Evolution

Evolution is change in the heritable characteristics of biological populations over successive generations.

These characteristics are the expressions of genes that are passed on from parent to offspring during reproduction. Different characteristics tend to exist within any given population as a result of mutation, genetic recombination and other sources of genetic variation. Evolution occurs when evolutionary processes such as natural selection (including sexual selection) and genetic drift act on this variation, resulting in certain characteristics becoming more common or rare within a population. It is this process of evolution that has given rise to biodiversity at every level of biological organisation, including the levels of species, individual organisms and molecules.

The scientific theory of evolution by natural selection was conceived independently by Charles Darwin and Alfred Russel Wallace in the mid-19th century and was set out in detail in Darwin's book On the Origin of Species. Evolution by natural selection was first demonstrated by the observation that more offspring are often produced than can possibly survive.

This is followed by three observable facts about living organisms: (1) traits vary among individuals with respect to their morphology, physiology and behaviour (phenotypic variation), (2) different traits confer different rates of survival and reproduction (differential fitness) and (3) traits can be passed from generation to generation (heritability of fitness).

Thus, in successive generations members of a population are more likely to be replaced by the progenies of parents with favourable characteristics that have enabled them to survive and reproduce in their respective environments.

In the early 20th century, other competing ideas of evolution such as mutationism and orthogenesis were refuted as the modern synthesis reconciled Darwinian evolution with classical genetics, which established adaptive evolution as being caused by natural selection acting on Mendelian genetic variation.

All life on Earth shares a last universal common ancestor (LUCA) that lived approximately 3.5–3.8 billion years ago.The fossil record includes a progression from early biogenic graphite, to microbial mat fossils, to fossilised multicellular organisms. Existing patterns of biodiversity have been shaped by repeated formations of new species (speciation), changes within species (anagenesis) and loss of species (extinction) throughout the evolutionary history of life on Earth. Morphological and biochemical traits are more similar among species that share a more recent common ancestor, and can be used to reconstruct phylogenetic trees.

Evolutionary biologists have continued to study various aspects of evolution by forming and testing hypotheses as well as constructing theories based on evidence from the field or laboratory and on data generated by the methods of mathematical and theoretical biology. Their discoveries have influenced not just the development of biology but numerous other scientific and industrial fields, including agriculture, medicine and computer science.

The Big Bang

The Big Bang theory is a cosmological model of the observable universe from the earliest known periods through its subsequent large-scale evolution. The model describes how the universe expanded from an initial state of extremely high density and high temperature, and offers a comprehensive explanation for a broad range of observed phenomena, including the abundance of light elements, the cosmic microwave background (CMB) radiation, and large-scale structure.

Crucially, the theory is compatible with Hubble's law – the observation that the farther away galaxies are, the faster they are moving away from Earth. Extrapolating this cosmic expansion backwards in time using the known laws of physics, the theory describes a high density state preceded by a singularity in which space and time lose meaning. There is no evidence of any phenomena prior to the singularity. Detailed measurements of the expansion rate of the universe place the Big Bang at around 13.8 billion years ago, which is thus considered the age of the universe.

After its initial expansion, the universe cooled sufficiently to allow the formation of subatomic particles, and later atoms. Giant clouds of these primordial elements – mostly hydrogen, with some helium and lithium – later coalesced through gravity, forming early stars and galaxies, the descendants of which are visible today. Besides these primordial building materials, astronomers observe the gravitational effects of an unknown dark matter surrounding galaxies. Most of the gravitational potential in the universe seems to be in this form, and the Big Bang theory and various observations indicate that it is not conventional baryonic matter that forms atoms. Measurements of the redshifts of supernovae indicate that the expansion of the universe is accelerating, an observation attributed to dark energy's existence.

Georges Lemaître first noted in 1927 that an expanding universe could be traced back in time to an originating single point, which he called the "primeval atom". For several decades, the scientific community was divided between supporters of the Big Bang and the rival steady-state model, but a wide range of empirical evidence has strongly favored the Big Bang, which is now universally accepted.[8] Edwin Hubble concluded from analysis of galactic redshifts in 1929 that galaxies are drifting apart; this is important observational evidence for an expanding universe. In 1964, the CMB was discovered, which was crucial evidence in favor of the hot Big Bang model,[9] since that theory predicted a uniform background radiation throughout the universe.

REFERENCES:

Books :

Book Name	The Authors	University	Date of Publication	CHAPTER/ PAGES
Programming the Universe	Seth Lloyd	The University Of Cambridge	5 April 2007	7/149
The Computing Universe: A Journey through a Revolution 1st Edition	by Tony Hey & Gyuri Pápay	The University Of Cambridge	8 September 2014	13/263

websites :

Web site name .	Day, month and year
Wikipédia https://ar.wikipedia.org/wiki/	07/05/2020
Syrian Researchers https://www.syr-res.com	17/05/2020
Al Khaleej Newspaper http://www.alkhaleej.ae	20/05/2020
The Huffpost Newspaper https://www.huffpost.com	27/05/2020

All quotes mentioned in the book are taken from this website : https://www.brainyquote.com/

INDEX

A

afterlife	116
Alan Turing	72
Albert Einstein	70
alchemical novel	87
Alfred North Whitehead	28
Al-Khwarizmi	27
Allen Newell	25
Alonzo Church	72
Alpha GO	24
Analysis of Logic	28
and Wallace, & Rabin	101
antibodies	34
anti-serum	34
Archimedes	70
Aristotle	27
Artificial intelligence	24
artificial intelligence technologies.	29
atoms	72, 164
automated or industrialized	27
awareness	52

B

bananas	63
Bergson	100
Bertrand Russell	28
Big Bang	86, 124, 162
binary system	29
Biocentrism	117, 156
biology	12

blood	35
Boolean algebra	28, 143
bottle got affected	111
brain	37
Broca region	41

C

capacitor	37, 140
Charles Babbage	78
Charles Eugene	126
chemical reactions	32
chromosomes	128
classic computer	76
colors	9
commutative law of addition	15
Computation	13
computational message	133
computing	71
computing and physics began in the early	78
concept of time	108
condenser	7
conscious	48
Consciousness is a consequence of the unconscious	48
correspondingly	105
Cosmological horizon	85
creative designer	125
creativity theory	125
Cunningham	101

D

Daedalus	26
dangerous diseases	32
Darwin	69
David Hilbert	28
death	117
delicious dish	63
dilemma	29
dimension Infinite	93
DNA	128
Dr. Burhev	131, 132
Dr. Robert Lanza	117
drugs	53

E

Edison Thomas	88
Edward Fredkin	74
Egypt (Pharaonic civilization).	26
electric generator	36
electrical circuit	34
electrical wires	9
electricity	8
Electroencephalogram	38
electromagnetic wave	8
electromagnetic wave of light	66
electromagnetic waves and transmission	8
Electromagnetism	138
electrons	72
Eliade	101
Emil Post	72
energy is unlimited	84
Euclid	27
Everything is a computation	69
Evolution	159
eyesight	66

F

Films	53

G

game of logic	109
genetic information and Biological	128
George Paul	28
glycoproteins	47
God	5, 94, 121, 124, 129
Google's	24
Gottfried Leibnitz	28
Gottlob Frege	28
Greek mythology	26
group of real functions	92

H

happy	120
heart	36
hegemony of the future	100
Heidegger	98
Hephaestus	25, 26
Hera	26
hero	53
Hilbert program	29
history of artificial intelligence	27
history of logic, arithmetic, and philosophy	27
human body	34
human brains	31
human intelligence	21
human senses	67
humanities discipline	21
hypothalamus	43, 44

I

image	9
immunological memory	34
inner mind	49
intelligent design	125
intensity	9
Internet	71

Isaac Asimov	78
Islamic civilization	27

J

Jabir bin Hayyan	27

K

Karl Wernke	41
Kelly	100
knowledge journey	11
Konrad Zuse	74
Kummel	101
Kurt Gödel	72

L

Laplace	77, 94
law of stone	17
lightning struck	32
limbic system	44
linear application	92
linguistic software	51
linguistic tasks	42
logic is only a calculation	27
logic of the human mind	30

M

Magnetoencephalography	39
Manfred Eigen	127
Margulis-Levitin theory	81
material universe	91
mathematical calculations and algorithms	23
mathematical formulas	70
mathematical laws	20
matrix	93
Max Planck Institute for Biochemistry	127
medications and treatments	33

memory	46
MR NOBODY movie	112
Muslims and Christians and Jews	116

N

native language	51
Neural signals	38
Neurologists	37
neurons	38
neurophysiological	39, 105
Nevertheless	66
Newton	69
Nikola Tesla	88
Nobel prize	88
Nobel Prize in Chemistry in 1968	127
nothingness	120

O

optical wavelength	9
our experience	61

P

pain program on the computer	10
Paolo Coelho	87
paranormal subconscious	49
people perceive	30
philosophical and artistic dialogues with humans	24
phone	9
Photons	9
physical and social aspect	107
physics laws	76
physiology	12
Piaget	101
Pierre Paul Broca	41
Pistachio Theory	111
Planck's constant	86

Principia Mathematica	28
Programming the Universe	71
protein to be created by chance	126
prove the existence of the computation using computation	14
psychological	107
publicity	53

Q

quantum computer	70, 76
quantum computer can simulate physical laws	76
quantum computing models	75
quantum dimension	118
Quantum mechanics	152
quantum systems,	75

R

radio	7
randomness and chance	125
read other people's ideas	49
read other people's thoughts	54
Reading	13
remote control	88
René Descartes	27
resurrection	116
robot	59
rock	71
Rolf Landauer	78

S

science of alchemy	27
sense of smell	65
sense of touch	66
série harmonica,	112
Seth Lloyd	71, 82

short story written by a Japanese artificial	25
small decision	111
smell of bacteria	64
Soil law	16
Spacetime	149
speech process	13
spirit of the universe	87
Stephen Hawking	123
Stephen Kleene	72
Stephen Wolfram	78
super laptop	79
supercomputer	81, 82, 83
superconducting quantum interfaces	40

T

tension	9
The ancient Greeks atom	77
The domination of the past	100
the five senses	67
The Last Question	78
the principle to read people's thoughts	58
theory of butterfly effect	111
theory of computation	155
theory of evolution	128
Thomas Hobbes	27
three-dimensional printer,	91
tomography	41
Turing	29
Turing machine	29, 144
Turing-Church hypothesis	73
TV	9

U

unconscious	48
unconsciously	42
unique person	54
universal rules of thought	30

universe is quantum		75

V

vector space		134
velocity average		81
viruses		32

W

Werner Heisenberg		118
Whitehead		100

Z

Zeus		26

Everything Is a Computation

www.ingramcontent.com/pod-product-compliance
Lightning Source LLC
Chambersburg PA
CBHW071125240526
45465CB00024B/1082